2-23-77

The New World Primates

Th

MARTIN MOYNIHAN

New World Primates

Adaptive Radiation and the Evolution
of Social Behavior, Languages, and Intelligence

PRINCETON UNIVERSITY PRESS · 1976

For my wife

1947541

Contents

Preface

There have been so many books on primates—monkeys and apes and their relatives—in recent years that one hesitates to produce yet another. It would seem that some justification is called for. Perhaps the only valid excuse is that primates are particularly interesting. They are, or should be, interesting for several reasons.

Man himself is a primate. He is descended from something that would be called an ape were it still living today. Apes are descended from animals that would be called monkeys were they still living today. One would like to know and understand the factors that have influenced or controlled the changes from one stage to another.

The monkeys of Central and South America may be relevant in this connection, although in a rather peculiar way. They are not very closely related to the immediate ancestry of man, which developed from one of the stocks of Old World primates. Certain anatomical details would indicate that they are less closely related to the monkeys and apes of the Old World than the latter are to one another. Yet they seem to have evolved many of the same characteristics independently, probably several times. They also are abundant and rather diverse. Thus, they provide unusually favorable opportunities for analysis of both evolutionary radiation, the appearance of new adaptations to new manners of life, and repeated convergence in similar physical and biological environments.

My qualifications for reviewing them consist of some sixteen years of intermittent observations of many species in captivity and in the wild. Six types were studied with some degree of thoroughness: *Saguinus geoffroyi, Cebuella pygmaea, Aotus trivirgatus, Pithecia monacha, Callicebus moloch,* and *Saimiri sciureus.* Additional forms of some of the same genera and one or more representatives of all the other genera (and a majority of distinctive subgenera) were observed more briefly. I have also relied upon the published reports of other students and verbal information from both professional scientists and non-technical observers in the field. I must thank these informants and hope that I have not distorted their accounts unduly.

Every author has preferences and biases. My own will be evident. I have been most concerned with patterns of social behavior and some interactions among individuals, and I have tried to trace their development in their historical and ecological contexts. This book is not meant to be a comprehensive or "balanced" summary of the whole of the biology or even the ethology of New World primates; rather it is a series of descriptions and discussions of special topics that seem to me to be significant, suggestive, or amusing.

The task of preparing and writing up the material was facilitated by much editorial, bibliographic, technical, and secretarial advice and assistance. I am particularly grateful to the secretaries, who must have been driven to distraction by the nature of the texts with which they had to work, and to Mrs. Alcira Mejía, Mrs. Bernadette French, and Mr. Jack Marquardt of the Smithsonian libraries.

Mr. John Hannon, of the Princeton University Press, was always encouraging as well as helpful.

The editors of the Smithsonian Institution Press and the *London Journal of Zoology* kindly gave permission to reproduce some of the figures.

I am indebted to the Smithsonian Institution itself for having provided me with so many opportunities for research.

Primatology is changing and progressing, and any review may become obsolete. I should, therefore, specify the time of my comments. A draft of this book was finished in September of 1973. Changes and additions were made later, in 1974.

MARTIN MOYNIHAN

Panama City, the Republic of Panama

The New World Primates

Infant Callimico goeldii.

chapter one
NOTES ON CLASSIFICATION AND HISTORY

The Order Primates

Most primates are easy to recognize as such, but the recognition usually is based upon unconscious assessment of data that are difficult to describe or summarize concisely.

The classical definition of the order was by Mivart (1873): "Unguiculate, claviculate, placental mammals, with orbits encircled by bone; three kinds of teeth, at least at one time of life; brain always with a posterior lobe and calcarine fissure; the innermost digit of at least one pair of extremities opposable; hallux with a flat nail or none; a well-developed caecum; penis pendulous; testes scrotal; always two pectoral mammae."

Unfortunately, this definition is neither all-inclusive nor very useful. Many characteristics of soft anatomy are not evident in most fossils. More important, there are several species, living or extinct, that have always been considered to be primates although they lack some of the characters cited by Mivart.

Another problem was noted by Huxley (1876). Surveying the primates as a whole, he said that "Perhaps no order of mammals presents us with so extraordinary a series of gradation as this—leading us insensibly from the crown and summit of the animal creation down to creatures from which there is but a step, as it seems, to the lowest, smallest, and least intelligent of the placental mammals."

Recent discussions have used more sophisticated terms, but are not necessarily more helpful. To mention one example, Martin (1968a) has suggested that primates can be identified by a rather recondite character of the bony skull, central nervous system, and reproductive organs. He is not certain, however, that any of these features are diagnostic per se. Moreover some of them would appear to bear little or no causal relation to the adaptations that have been responsible for the greatest successes of the order.

It may be prudent, therefore, to be vague. For most practical purposes, primates can be described simply as placental mammals primarily adapted to arboreal life, usually eating a variety of foods, and with more or less grasping hands and feet, large and complex brains, and binocular vision.

Even this must be qualified by exceptions. There are primates that spend (or may be supposed to have spent) all or most of their time on the ground. But they show unmistakable signs of being related to or descended from (sic) arboreal forms.

Whatever the difficulties, most people would agree that the living primates include the following:

1. Lemuroids. Many species of the island of Madagascar, the "true" lemurs and some more exotic animals.

2. Lorisoids. The lorises of southern Asia and the galagos and pottos of continental Africa.

3. Tarsioids. Several species of the genus *Tarsius*, confined to the Philippine and other East Indian islands.

4. Ceboids. The native primates of the New World, apart from man.

5. Cercopithecoids. The guenons, mangabeys, macaques, baboons, langurs, guerezas, etc. of the Old World. They range through most of continental Africa, large parts of Asia, and some nearby areas such as Gibraltar and Celebes.

6. Hominoids. Man and the "great" apes, including the gibbons, Siamang, and Orang-utan of southeast Asia, and the chimpanzees and Gorilla of Africa.

These groups have been assigned different ranks, arranged in different ways, by different specialists. Simons (1972) provides an authoritative recent classification, to the generic level, of both living and fossil primates, and also cites checklists by other authors. Two major supergroups are obvious.

The lemuroids, lorisoids, and tarsioids appear to be less specialized or "progressive" than the other living primates in some respects. With their fossil relatives, they usually are placed together (following Simpson, 1945) in a separate suborder, the Prosimii, and given the "vernacular" name of prosimians.

The other groups are usually bracketed in a suborder Anthropoidea. Most of the ceboids and cercopithecoids are called monkeys. By "monkey" the layman usually means a moderately large primate, comparatively intelligent, disconcertingly manlike in some aspects of behavior, but with a visible tail. This is as good a definition as any, although somewhat misleading in a few special cases. By "ape" the layman who speaks a language, such as English, that distinguishes among different types of primates usually means something very like a monkey, but probably even larger and certainly without a conspicuous tail. Biologists who use the term usually restrict it to Hominoids apart from man.

Terms such as "intelligent" and "intelligence" are also difficult to define. I shall not attempt to analyze the various concepts that might be involved. It will be sufficient to say that the terms are being used in the ordinary, everyday sense. In effect, this means that intelligence is equated with ability to learn new things and to apprehend relations between perceived phenomena. (For a discussion and summary of some of the ways in which the term "learning" has been used, see Thorpe, 1956.)

Of course, it is impossible to test the intelligence of extinct animals. And really very little is known of the learning abilities of some of the living primates. But this is not quite as serious a handicap as might be supposed. Among those animals whose behavior has been studied in the laboratory or in other controlled situations, there has been found to be a general positive correlation between size and superficial complexity of brain and intelligence. Small animals usually have relatively larger brains than otherwise similar large animals. Among species of similar body size, however, the forms with larger brains usually prove to be more intelligent than those with smaller brains. At any given size, mammals whose cerebral hemispheres are convoluted on the surface tend to be more intelligent than those with smooth brains; and convolutions may make impressions upon the inner walls of the cranium. Thus morphology alone, the size, shape, and other aspects of the skull, can be used as a crude gauge of intelligence. (Comparative aspects of relative brain size and development are discussed in Rensch, 1956 and 1960, Bauchot and Stephan, 1966, Stephan, 1967 and 1972, and Jerison, 1973.)

The Earliest Primates

As indicated above, some of the primates are quite similar in basic anatomy to the most primitive of placental mammals. The latter are conventionally placed in the order Insectivora. The tree shrews (family Tupaiidae) would seem to be the most nearly primitive of living Insectivora. It is not coincidental that they were also at one time considered to be the most primitive of living primates. This opinion is now unfashionable (see, for instance, Van Valen, 1965, Campbell, 1968, and Martin, 1968b), but it seems probable that the first primates cannot have been very different from tupaiids in either behavior or ecology.

Living tree shrews are confined to tropical Asia. They are small animals of not very distinctive appearance (remotely similar to squirrels, ordinary shrews, or opossums) with rather small brains. They have been studied almost exclusively in captivity, but thoroughly enough to give some idea of their normal activities. The best known forms, possibly all subspecies of *Tupaia glis*, seem to be diurnal, nearly omnivorous, and partly terrestrial and partly arboreal (Kaufmann, 1965, and Martin, 1968b). They use their hands for grooming and occasionally holding food, despite a primitive arrangement of fingers that is not particularly suited to the more complex or delicate kinds of manipulation (Bishop, 1962 and 1964).

Adults are not very gregarious, but they have a fairly extensive repertory of social signals. They seem to have more different types of patterns that have become specialized for communication than other shrews, but less than all or most typical primates (Moynihan, 1970a). Many other species of the family Tupaiidae seem to be essentially similar. A few may be slightly more highly social, more frequently or extensively gregarious (Sorenson and Conaway, 1966).

Starting from some such source, the true primates developed at the end of the Cretaceous period and the beginning of the Tertiary period, approximately 70 to 65 million years ago (Van Valen and Sloan, 1965, McKenna, 1966, Szalay, 1967, Simons, 1963, 1964, 1972). Presumably the adaptive shift involved an increase of arboreality and a greater reliance upon vegetable foods. (The precise sequence of changes is not entirely clear, and there may have been additional factors at work—see Cartmill, 1974a—but the general trend is unmistakable.) All the earlier primates can be called prosimian. They seem to have become extremely varied between the middle Paleocene and late Eocene epochs, perhaps 60 to 34 million years ago. Many of the Paleocene types were rather surprisingly rodentlike, even more so than tree shrews, with procumbent and gnawing front teeth, but most of the better known Eocene species were more nearly comparable to the modern lemuroids and tarsioids.

Of the surviving prosimians, only the lemuroids are flourishing enough to suggest the probable range of habits of their earlier relatives. They will be described in some detail in Chapter 6, but a few points may be noted here in anticipation. The living species are both more varied than tree shrews and more specialized on the average. They are small to medium large in size. Some are diurnal. Others are nocturnal. They all eat vegetable matter. A few also take insects and other small arthropods, which they may catch in different ways. Some species and subspecies are highly gregarious, living in large bands that include several adults of both sexes. Others live in apparently stable family groups of one adult male and one adult female with their most recent young. Still others are often solitary. All of them have more or less elaborate signal systems. The great majority of the species are arboreal. They have hands and feet that are better adapted to grasping than are those of tree shrews, but they seem to use the increased capability primarily for locomotion, seizing trunks and branches of trees and bushes, rather than for manipulation of other objects (A. Jolly, 1966).

Although the brains of most living lemuroids and tarsioids are proportionately smaller and apparently simpler than those of living monkeys, they are more highly developed than those of tree shrews and they are not relatively small in comparison with the brains of most living mammals of other orders. What little evidence there is indicates that Eocene prosimians had brains like those of their present day counterparts (perhaps with more emphasis on olfaction, the sense of smell, in some cases). As most other Eocene mammals had much smaller brains than their nearest living relatives, this would seem to mean that the early primates were already the most intelligent of contemporary animals. See also Radinsky (1970). It is interesting, and must be significant, that the primates had already acquired this advance or advantage at the prosimian stage.

The Eocene prosimians, in turn, must have given rise to all the "higher" monkeys and apes, including the forms that are now found in Central and South America.

The Ceboids

The monkeys of the New World have often been divided among a host of taxonomic categories, families, subfamilies, tribes, etc., with a corresponding proliferation of complicated and unstable names. Some of this effort seems to have been sheer embroidery or hairsplitting. The living forms fall naturally into some eight or nine subgroups. I would suggest that they be listed as follows:

1. Tamarins. Genera *Callimico, Leontopithecus,* and *Saguinus.*

2. Marmosets. *Callithrix* and *Cebuella.*

3. The Night Monkey or Douroucouli. *Aotus.*

4. Howler monkeys. *Alouatta.*

5. Sakis and uakaris. *Pithecia* (including *"Chiropotes"* and *"Cacajao"*).

6. Titi monkeys. *Callicebus.*

7. Spider monkeys and the Woolly Monkey. *Ateles* (including *"Brachyteles"*) and *Lagothrix.*

8. The Squirrel Monkey. *Saimiri.*

9. Capuchin monkeys. *Cebus.*

Most of the scientific names in this list conform to Hershkovitz (see below). Names in quotation marks are traditional in the literature, but the taxa to which they have been applied do not, in my opinion, deserve to be recognized as separate genera. Many resemblances between taxa can be conveniently shown by setting broad, inclusive generic limits. The classification proposed and used here favors "lumping" over splitting as a general rule.

The limits and contents of the subgroups have been determined by studies of extant animals. In the circumstances, it is impossible to arrange them in any clear linear sequence or hierarchy. With one or two exceptions, each subgroup appears to be almost equally distinct from every other. Similarities and differences are distributed in complex patterns.

It would also be difficult to establish that any one of the subgroups is conspicuously more primitive on the average than several others. Most of them exhibit an array of both obviously specialized and primitive features of morphology and behavior. (Hershkovitz, e.g. 1970 and 1968, suggests that the small tamarins and marmosets are primitive among the living forms, and that the smallest, *Cebuella,* is the most primitive of all. This is debatable. It is true that many early primates were small. But the size of tamarins and marmosets, and many other characters that seem to be functionally related, are directly adaptive. They could be "secondary" specializations. They certainly are combined with progressive features.)

There are enough resemblances among ceboids to indicate that they are monophyletic. The immediate common ancestor is not known, or at least has not been recognized and definitely identified as such. It may have already advanced beyond the prosimian stage toward the monkey level.

The courses of evolution of particular lineages of New World primates are obscure. They do not, as monkeys, have a good fossil record. The earliest traces of unmistakable ceboids are of Oligocene and Miocene ages, some 34 to 12 million years ago, and were found in South America. South America was an island during all or most of the Tertiary, and the home of a most peculiar fauna of mammals, birds, and other vertebrates (Darlington, 1957, and Patterson and Pascual, 1968 and 1972), some of which have survived to the present day, others of which have not. The ceboids were and are among the diagnostic members of this fauna. They would appear, however, to have reached South America, or expanded within it, at a later date than some other kinds of mammals (see Hoffstetter, 1972, as well as Patterson and Pascual). The published accounts of the Oligocene and Miocene types are difficult to interpret and to reconcile with one another. Stirton (1951) thought that these animals were near to some of the living forms. Hershkovitz (1970) suggests that most of them were more distinctive, representatives of lines that have not survived. In any case, it is evident that the ceboids had become diversified by the end of the Miocene. The remains of monkeys discovered in late Pleistocene or younger deposits on the mainland of South America, only a few thousands, or tens of thousands, of years old, are nearly or completely indistinguishable from the corresponding parts of living animals.

A few ceboids occur in Central America, essentially the tropical part of North America, at the present time. Tropical climates and habitats must have been more extensive in the region during some earlier periods. Doubtless they would have been suitable for monkeylike primates. An enigmatic fossil monkey, *Xenothrix*, has been found in Jamaica (Williams and Koopman, 1952). Apart from this, there are no indications that Central America or tropical North America could have been the site of a major or independent evolution of ceboid groups. The monkeys that occur there now are recent immigrants (see also below). South America is the home of the successful ceboids, and their basic adaptations are to South American conditions.

chapter two

THE SETTING

The physiography of South America is simple in broad outline, with the high chains of the Andes to the west, older and lower mountains in parts of the east, and extensive lowlands in the drainage basins of the great rivers, the Magdalena, Orinoco, Amazon, Paraná, and others. The area is approximately 11,200 square kilometers. A very substantial proportion, perhaps 75 per cent, lies between the tropics of Capricorn and Cancer. (Figures taken from Keast, 1972a.)

The bulk of the tropical area is lowland. At the present time, much of it is covered by different kinds of forest and scrub, including rain forest. Under "natural conditions," that is, present climates and no interference by man, the areas of forest and scrub in general, and rain forest in particular, would be considerably larger. Rain forest is the richest and most diverse of plant formations in the sense of being composed of a multitude of species of very different shapes and sizes (Richards, 1952 and 1973b, and Parsons and Cameron, 1974). As would be expected, the fauna that lives there is also exceedingly diverse, more so than in any other terrestrial habitat.

This system probably has existed in South America for a very long time, although not without variations. There certainly have been climatic vicissitudes, alternate periods of relatively greater and lesser humidity, especially during the Pleistocene (see summaries and references in Simpson Vuilleumier, 1971, Vanzolini, 1973, and Raven and Axelrod, 1974). Students of birds (Haffer, 1967, 1969, and 1974) and lizards (Vanzolini and Williams, 1970) have argued that the humid forest must have been repeatedly broken up into separate pockets or patches, "refugia," during drier periods. They have also suggested that the process contributed to the multiplication of species in the tropics. Originally similar populations of forest organisms may have become differentiated during the stages of fragmentation. They could have developed divergent ecological preferences and potential reproductive isolating mechanisms. These, in turn, might have permitted their descendants to coexist without interbreeding when the various patches spread and coalesced again during wetter periods. See also Mayr (1969).

This phenomenon would seem to be real; but its effects may have been limited or small scale in the lowlands of tropical South America. The best evidence that forests have been predominant throughout the history of tropical South America is that most of its fauna is adapted to forested conditions. It includes relatively very few open country or savannah types (Vesey-Fitzgerald, 1964, and Hershkovitz, 1969 and 1972).

There are savannahs in parts of lowland, tropical South America now. Their origin is a matter of some dispute (points of view are presented in Beard, 1953, Goulissachvili, 1964, Talbot, 1964, and Vesey-Fitzgerald). It seems probable that most of them are artificial, the products of comparatively very recent human activities.

Tropical America would seem to have escaped the really severe and widespread aridity that has been characteristic of some parts of the Old World tropics, most notably Africa (Moreau, 1966), during the last million years or so. This may be one of the reasons why its flora is richer than that of tropical Africa (Richards, 1973a). It might also help to explain some of the differences between the ceboids and their Old World relatives (see below).

American monkeys do not betray the effects of having been confined to refugia in the same way or to the same extent as some birds. Different genera of ceboids have different ranges. Some of them include many more visibly distinct forms than do others. In most cases, however, each genus is represented by no more than a single type in any given area. Many other kinds of mammals illustrate the same general rule. Extensive sympatry, i.e., overlapping of ranges, of closely related forms seems to be much rarer among the larger nonflying mammals of Central and South America (everything "above," in several senses, the rats and mice) than among the birds of the same areas or comparable mammals of other continental masses of similar latitudes, such as Africa and southern Asia. This must mean that they have had fewer opportunities to speciate in geographic isolation. (Possibly the periods of reduction and shattering of rain forests were only brief episodes in the New World tropics and/or the shattered fragments were still connected by corridors of more open forest or scrub that were not prohibitive barriers to larger mammals, whatever their effects upon other organisms.)

There must also have been biological and zoogeographical changes of different magnitudes and on very different time scales.

It has often been assumed that the tropics are "stable." The statement is misleading as a generalization, although not always in detail. There are appreciable fluctuations in population numbers and distributions from year to year and decade to decade in humid forests and elsewhere (personal observation—see also comments in such works as Eisenberg and Thorington, 1973). It is even possible that changes in tropical biotas and environments are more unpredictable in the short run or in some respects than are those of the north temperate zone. May (1972 and 1973) discusses, with a wealth of mathematical formulas, some of the relations and conditions that might favor instability in complex systems. To put it crudely, the choice is between correction by homeostasis, feedback mechanisms, and disturbance by domino effects in chain. Obviously both have occurred (MacArthur, 1972). More important, in fact, are the unmistakable and undoubted revolutions that have succeeded, as connections among land masses, plates, and continents, have shifted over the longest relevant time span, the last few tens of millions of years.

The present faunas of South America, Central America, and the West Indies are mixtures. They include elements that evolved in North America and in South America when the two continents were separated during the Tertiary and that later strayed out to the islands or migrated north or south along the mainland when a connection was (re)established in the vicinity of the isthmus of Panama and northern Colombia in the late Pliocene or early Pleistocene. Today the three areas and the various mixtures are usually considered by zoogeographers and other biologists to form a single "neotropical region." All the warm, lowland continental and isthmian parts can be included in a single subregion. This may be called "Brazilian," following Hershkovitz (1958). The most conspicuous division within it, at the present time, is imposed by the Andes. These separate some of the organisms of the Amazon and Orinoco basins from those of Central America and the Pacific coast of Colombia and Ecuador. But the mountains have not been a complete barrier, and many of the animals and plants are quite similar on both sides. Only the biota of the upper Amazon valley is somewhat more variegated, possibly because physical conditions are somewhat more favorable there.

The actual components of the modern neotropical mammalian fauna are described by Hershkovitz (1972), summarized and analyzed by Keast (1972b), and also by Eisenberg and Thorington (op. cit.). They include 12 orders (perhaps 24 taxa at the combined ordinal and superfamily grades), 50 families, 278 genera, and approximately 810 species. (The figures for genera and species may well be overestimates.) The dominant groups, in terms of numbers of species, biomasses, and trophic levels, are monkeys, bats (Chiroptera), rodents (cricetine rats and mice, and the caviomorphs, some of which are ungulatelike, others of which are tree-living), opossums (didelphid marsupials, often arboreal), edentates (tree sloths, anteaters, armadillos), placental Carnivora (also often arboreal), and real ungulates (deer, tapirs, and peccaries in the tropical parts of the region). The mammals that are most likely to impinge upon monkeys as possible competitors are some of the opossums, the tree sloths, and some of the rodents, squirrels, and porcupines.

It may be revealing to consider one of the past cases of replacement in the neotropics. Placental carnivores are occasional predators of at least a few of the New World primates at the present time. This is interesting because the only mammalian predators in South America during most of the Tertiary belonged to a different and now extinct group, the borhyaenid marsupials. It is remarkable that the ceboids have been able to withstand an extensive changeover of a large series of predators (and not all that long ago) without more apparent damage. Although predation may be important in particular instances, and must have been one of the factors affecting the evolution of ceboids, it may not usually be the principal cause of mortality among them. All or almost all of the extant species take precautions against predators, sometimes elaborate precautions that are exceedingly visible to the biologist, but the majority of individuals may die from less visible causes, such as old age, malnutrition, and disease. (There are indications that predation is often *relatively* less important in the tropics than in some other regions—Moynihan, 1971.)

Other arboreal herbivores and omnivores, such as tree sloths and some opossums and rodents, would seem to have withstood the same change of predators with equal success. Perhaps some aspects of the arboreal habitus and habitats are as nearly imperturbable as other parts of the tropics are supposed to be?

Conservative or not, primates and other mammals have always had to cope with other organisms of different classes and even phyla. Birds are one example; South America has a superabundance of them. De Schauensee (1970) recognizes 93 families, 865 genera, and 2926 species. This is by far the richest assemblage of birds in the world (it is approximately twice the size of the avifauna of Africa). Much of it is tropical and occurs in forests. A few species of birds are predators

upon monkeys. Many others compete with monkeys for resources, food, resting and sleeping sites, observation posts, etc.
A not inconsiderable number are witting or unwitting collaborators of monkeys. And the monkeys, of course, may interact with the same or different species of birds in reciprocal or
reverse ways. The birds must both preoccupy resources that
would otherwise be available to the monkeys and provide
other resources that would otherwise be unavailable. Reptiles
may be almost as numerous and significant as birds in the
tropics. Arthropods, especially insects, are ubiquitous and
probably overwhelmingly more important as prey, predators
(parasites and vectors), and competitors than all the vertebrates put together. Whatever their past or present successes,
the ceboids can never have had more than a small piece of
the action in their environments.

The relevant factors are clear. The fauna of tropical America
as a whole is predominantly forest-loving, flying, and/or arboreal (actually tree-climbing). The situation is not new. It
must have set the limits within which the evolution of the
local monkeys has had to operate.

chapter three
NATURAL HISTORY

This chapter should be introduced with a warning. The information available on living ceboids is heterogeneous. Certain kinds of ecological and demographic data are complete, quantitative, and elegant. Other material is partial and anecdotal. It is all accumulating rapidly. Something is known of the habitat preferences, movements, and feeding habits of at least one member of each of the major subgroups. In some cases, it is even possible to compare several forms to estimate the range of variation within a subgroup.

The following accounts will be more concerned with genera, populations, and individuals than with the taxonomic units of species and subspecies.

Much of the narrative may appear to be tedious or repetitive and weakened by the constant use of weasel words such as "possibly," "probably," and "presumably." It is drawn to considerable length for two reasons: to permit an independent assessment of the reliability and consistency of the data (some of which are new), and to provide the necessary background material for the subsequent chapters. The reader who is interested in general questions and problems may want to skip this chapter at first, or skim through it lightly, and then refer back when necessary. There is also a brief and partial summary in the "Review" that follows Chapter 5.

Tamarins (genera *Callimico, Leontopithecus,* and *Saguinus*)

These are small monkeys. All or most of them eat many animal and vegetable foods in the wild, but seem to prefer insects and other small arthropods whenever they can get them. They certainly spend considerable time searching for insects under natural conditions. In captivity, they show a greater preference for animal foods than do some other types of New World monkeys such as *Callicebus moloch, Pithecia monacha,* and *Alouatta villosa.* (This may or may not be the result of a reversal in the course of evolution—see above.) There is a causal link between small size and insectivorous habits. Insects themselves are small and sometimes difficult to find or catch. They cannot usually provide sufficient nourishment for a large animal, or only some of them can do so in special circumstances. (Forms such as the Giant Anteater, *Myrmecophaga tridactyla,* depend upon insects that are colonial and, therefore, more concentrated than most.)

Other distinctive features of tamarins may be correlated with, perhaps consequences of, small size and insectivory. Thus, for instance, they have sharp claws on their fingers and on all their toes except the hallux, the homologue of the human big toe, unlike most other monkeys, which have flat nails instead. The claws afford purchase on branches and tree trunks. Tamarins are so light in weight that the purchase is usually adequate. They may also "need" claws more than many of their relatives because their hands and feet are too small to allow them to grasp sizeable branches between opposed digits in the same way as larger species. (This does not mean that they have absolutely no powers of grasping. The hallux is set off from, and can be opposed to, their other toes. Thus, their feet can be used to seize thin branches. But their fingers are arranged more or less radially, with some variations—see references in Cartmill, 1974b—and are only partly opposable to one another. Their thumbs are not noticeably more opposable than the other fingers. They do not, therefore, manipulate objects particularly frequently or expertly; perhaps no more than do lemurs.) Their cerebral hemispheres are also comparatively smooth. This may reflect the fact that they are not among the most intelligent of monkeys (although, in my experience, they are much more alert and vivacious than most lemurs in many aspects of social and maintenance behavior), but it is another general rule that small species of all groups tend to have

Figure 1. An adult Saguinus geoffroyi *leaping between branches. The tail is being used as a brake and a counterweight. The hairs of the distal part of the tail are raised in a hostile display (largely alarm).*

smoother hemispheres than related species of larger size. The members of the genera *Leontopithecus* and *Saguinus* differ from most other New World primates in having lost the third molar teeth in both upper and lower jaws, possibly because the jaws are too small and short to accommodate a full series of teeth of optimal individual size, possibly because arthropod prey can be triturated by a shorter or simpler battery of teeth than is necessary for the mastication of the tougher fibers and other difficult materials of many plant foods. (The equally small or smaller marmosets have suffered a similar loss. Some other insectivorous mammals, such as some of the Carnivora, have reduced or abbreviated dentitions—Ewer, 1973.)

Saguinus and *Leontopithecus* are very closely related to one another, very similar in most physical characters, not only their teeth, claws, and brains.

Callimico is somewhat apart (it has been placed in a separate taxonomic family or subfamily by many authors). It is slightly larger, and has retained the third molars. The latter is a primitive feature; the former may be also. *Callimico* further resembles the larger ceboids in bearing only one young at a time, unlike both the (other) tamarins and the marmosets, which usually give birth to twins. This would not appear to be of profound phylogenetic significance. Obviously related species of other groups of mammals may have different litter sizes. At least one form of *Saguinus, geoffroyi* (Moynihan, 1970b), usually raises only one of the two young born in most years. In appearance and actions in life, *Callimico* is very reminiscent of *Saguinus* and especially *Leontopithecus* (see page 2 and Figures 1, 2, and 3). It probably is a perfectly good tamarin, although doubtless peculiar in some respects.

As suggested by the figures, tamarins look like squirrels but have the faces of monkeys or old men and long, thin tails. The various species are extremely diverse in color and pattern. A few are essentially unicolored, dark (blackish) or light (white or golden yellow). The majority have different, sometimes very intricate, arrangements of black, white, rufous, and

Figure 2. Various adult tamarins. From top to bottom: Leontopithecus rosalia, Saguinus fuscicollis, S. oedipus, Callimico goeldii, Leontopithecus rosalia *again.*

Figure 3. Copulation by Saguinus geof-froyi. *The Coiling of the tail is a sexual display. After Moynihan (1970b).*

gray or yellowish tones. Intricacy of pattern may subserve several functions. It may be protective, a kind of "disruptive" coloration, in some situations and against some backgrounds. It may also increase conspicuousness and lend emphasis to social displays in other circumstances. Many species have elongated tufts and ruffs of hair around the face or head. These can be raised or depressed, singly or in combination, according to the mood of the animal, and they are important in visual signalling. See below, Chapter 5.

The genus *Callimico* includes a single species, *goeldii*. It occurs in the upper part of the Amazon basin and was discovered rather late in the history of the exploration of tropical America. The few early examples whose provenance was known came from southeastern Peru and adjacent areas (references in Hill, 1957). Recently, however, the species has been found far to the north, in the *intendencia* of the Putumayo in Colombia. I observed several individuals near the town of Puerto Umbría, approximately halfway between the cities of Mocoa and Puerto Asís (see map), about 300 meters above sea level; T. Whittemore (personal communication) extended these observations; H. Le Nestour (personal communication) saw other individuals in nearby areas; and Hernández-Camacho and Cooper (in press) and I were able to gather some additional information on the habits of the species from local settlers, hunters, and Indians.

All the evidence suggests that these animals are never very common but are spread through many different habitats. The local people in the Putumayo say that they can be found in both scrub (*rastrojo*) and forest (*monte grueso*), both on hills and in valleys. The individuals that I observed myself were living in mixed forest and scrub, mostly low and young second growth, possibly 5 to 10 years old, on a poorly drained island (flooded at frequent but irregular intervals) in the Rio Guineo. This is a region of high rainfall and atmospheric humidity. The vegetation of the island is extremely dense and rather varied for its apparent average age. It is further diversi-

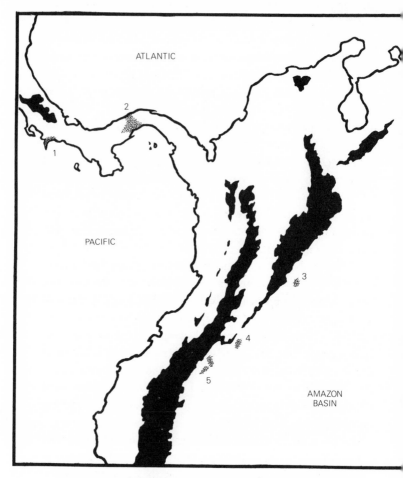

PANAMA AND COLOMBIA.

*The principal mountain masses and chains (approximately
2000 m. and above) are shown in black. The areas in which
I made most of my own observations are indicated by stip-
pling: 1, the Burica Peninsula and the neighborhood of Puerto Armuelles; 2, central Panama; 3, areas around San Martín
and Barbascal in the Meta; 4, neighborhood of Valparaíso in
the Caquetá; 5, scattered sites (El Pepino, Rumiyaco, Puerto
Umbría, Santa Rosa) in the Putumayo.*

fied or interrupted by a few large native trees, presumably relicts of an earlier forest, many stands of imported Asiatic giant bamboos, and occasional plantations of maize, fruit trees, and other crops. (The upper Amazonian region has not been turned into savannah, but neither has it remained immune to man. Parts of it have become almost suburban, rather less neat and clean than Surrey or Bucks County but quite comparable to New Jersey. And it has the beginnings of industrial pollution. Some kinds of human interference with the environment have been disastrous for monkeys. Others have been favorable, especially to the smaller forms.) The individuals observed by Le Nestour were found in mature forests that included many old and tall trees (probably rain forest or something very similar). This vegetation is much less dense than either the second growth and bamboo of the island in the Rio Guineo or the thickets and tangles of some of the habitats preferred by populations of *Saguinus* species (see below). Wherever they occur, individuals of *Callimico* wander through all levels of the available trees and bushes. They may tend to remain in the canopy when undisturbed, but they often bolt for the ground when surprised by potential predators, even large, awkward, and terrestrial animals such as man. Local Indians take advantage of this behavior by running them down with dogs. (One would suppose that much of the escape behavior of *Callimico* is designed to provide protection against flying birds of prey. It is not always entirely conventional. Two pairs that I surprised by an unexpected approach first stashed their infants in the nearest convenient low bush and then ran away through the trees. I have never seen any other American monkeys "abandon" their young in this fashion.)

Individual *Callimico* can walk, gallop, and leap quadrupedally and more or less horizontally like squirrels. They tend to leap more frequently and for longer distances than do the squirrels of similar size in the same and adjacent areas. When they perform such movements, they could be called "springers" in the sense of the term used by Erikson (1963). But they also cling to tree trunks in a vertical position, and often leap or hop from trunk to trunk with the body and head kept upright. This may be particularly likely to occur when they are frightened, if they cannot or do not choose to come down to the ground for one reason or another. Such vertical clinging and leaping is reminiscent of some prosimians, e.g., *Tarsius* and *Lepilemur* and the living members of the family Indriidae among the lemuroids. Napier and Walker (1967) suggest that it is primitive in primates. (This suggestion has not been accepted by all students, but the habit is at least old in some cases.)

Le Nestour saw *Callimico* eating insects and berries like other tamarins. I did not see any feeding myself. The owners of farms near Puerto Umbría say that individuals of the species are fond of the fruits of *Inga* and *Coussarea*, which are abundant in second growth, and they also raid plantations to eat the fruits of Cacao (*Theobromo cacao*) and "caimé," one of the cultivated Sapotaceae. (Most of the technical names of Colombian plants in this account are from Perez-Arbelaez, 1956, unless specifically stated otherwise.) R. Lorenz found that captive individuals would eat lizards, frogs, and even small snakes. Nevertheless, there are perhaps some slight indications that *Callimico* may take more vegetable matter than most of the other tamarins under natural or seminatural conditions. If so, this would help to explain some of its dental distinctions.

The genus *Leontopithecus* includes a single species, *rosalia*, with several well marked subspecies. They inhabit the humid coastal forests of southeastern Brazil (sometimes called "tupí" forests). These are being rapidly destroyed. Most of their native primates are becoming shy and rare. Recent observations of the remaining populations of *Leontopithecus* in the wild are conveniently summarized in Coimbra-Filho and Mittermeier (1973a). Individuals feed on many different kinds of insects, Blattariae, Orthoptera, Homoptera, Lepidoptera, Coleoptera (especially larvae), spiders and araneomorphs of the families

Lycosidae and Ctenidae, small lizards, birds' eggs, snails and earthworms (at least in captivity), buds, and fruits, especially those that are soft, sweet, and pulpy, e.g., *Tapirira guianensis, Inga* spp., *Cecropia* spp., bananas, and many forms of Myrtaceae. Some foods may be obtained by peculiar techniques. Hershkovitz (1972) notes that *Leontopithecus* has long slender arms, narrow palms, and elongated middle fingers. He suggests (without citing his sources) that these features are adaptations for digging insects and grubs from under loose bark and from cracks or holes in tree trunks and branches, like the similar but more exaggerated specializations of one of the marsupial phalangers of New Guinea, *Dactylopsila*, and the lemuroid aye-aye, *Daubentonia* (see also Chapter 6).

Saguinus is much more successful than either *Callimico* or *Leontopithecus* at the present time. It may, in fact, be the most successful of all the genera of ceboids. It includes more species than any other genus. It is widely distributed throughout most of the Amazon basin, the coasts of Colombia, and southern Central America, perhaps as far as the frontier between Panama and Costa Rica, but it does not extend to southeastern Brazil. Many or all of the species are still abundant. A few have been observed in detail and at length. The two best known, *geoffroyi* and *fuscicollis*, differ in minor aspects of anatomy, such as distribution of hair on the head and face and general color tone and pattern, and belong to different species groups. Both are somewhat marginal within the range of the genus as a whole, but they probably are fairly representative in ecology.

S. geoffroyi lives in Panama and northern Colombia. In Panama, it is most numerous in dense but not very tall forests or high scrub on well drained ground at low elevations.

Vegetation of an appropriate physiognomy may be "climax," the end product of undisturbed growth, in areas where there is an alternation of wet and moderately severe dry seasons, especially on rocky or other poor soils (see Figures 4 and 5). It is possible that dry and comparatively low forests were nearly continuous along most of the Pacific coast of Panama during some of the post-Pleistocene or sub-Recent a few thousands of years ago, before the arrival of man in the region. If so, the distribution of the tamarins may have been equally continuous in the same areas at the same time.

S. geoffroyi does not do well—perhaps it cannot survive for long—in the taller but less dense forests that are climax in more consistently humid areas with little or no dry season. It is not even common in or around the small windfalls, which are always scattered about. There is reason to believe that tall humid forests were nearly continuous along the Atlantic coast of Panama a few thousand years ago. If so, the tamarins may have been almost or entirely absent from the area.

Of course, all this has changed. The effects of human activities upon the tamarins have been marked. They may be worth describing because they are very mixed and perhaps typical in their ambivalence. (The *S. geoffroyi* case is clearer than that of *Callimico*.) The crucial factor would seem to have been the "slash-and-burn" agricultural techniques of the pre-Columbian Indians, their colonial successors, and many modern farmers. This entails clearing a patch of forest or scrub, raising crops on the cleared land for a few years, seldom more than three, and then moving on to another site, leaving the first patch to be overgrown by new scrub and second growth forest. *S. geoffroyi* individuals cannot remain permanently on cleared land, although they may visit plantations; but the second growth that develops on the Pacific coast of Panama quickly becomes suitable for them. More remarkable, so does some of the second growth, following human clearing, on the Atlantic side. (Much of the "artificial" second growth on both sides would appear to be structurally similar to the supposedly original, primary forest of the Pacific coast, even if the species composition is different.) Abandoned crop fields are much larger than windfalls on the average. Presumably as a result, *S. geoffroyi* has been able to invade and proliferate in some

Figure 4. Low and dense (possibly climax) forest on Ancon Hill on the Pacific coast of Panama. Note how closely everything is crowded together. Photo by O. Linares.

areas of the Atlantic coast of Panama that were previously
unavailable to it, most notably near the city of Colón and
along the borders of the Canal Zone, where there are many
abandoned fields, fairly close together.

Thus, the human population of the isthmus has succeeded
in changing the distribution of the tamarins in two ways,
making it both patchier and wider than before (see also Moynihan, 1970b).

It would not be an exaggeration to state that *geoffroyi* has
become a real commensal of man in Panama. (This is probably
true in Colombia, too. Hernández-Camacho and Cooper, op.
cit., note that the local individuals of the species are most
abundant in secondary forests or early secondary growth interspersed with plantations.)

In all the Panamanian areas inhabited, tamarins usually
prefer to remain near, but not exactly at, the edges (top and
sides) of the forest and scrub. These places are characterized
by many tangles of vines and lianas, sometimes substantial in

*Figure 5. Another aspect of low and dense
forest on Ancon Hill. Photo by O. Linares.*

mass but often rather delicate in texture with a relatively high proportion of small and thin tree trunks and branches.

C. M. Hladik et al. (1971) list some of the foods taken by the tamarins of Barro Colorado Island in the Canal Zone of central Panama, and estimate the nutritive contents of these materials. Protids comprise 20.6% (16.0% of animal origin), lipids 9.1%, reducing glucids 29.0%, cellulose 7.3%, and "complementary fractions" 34.0%. The significant item is the relatively high proportion of animal matter, higher than in the diets of the other New World primates that have been subjected to the same sort of analysis. Unfortunately, there are problems of interpretation. As the authors stress, the data are not numerous and involve considerable extrapolation. Barro Colorado, at the moment, is covered by humid "monsoon" forest, once cleared but now apparently approaching maturity again (Bennett, 1963). The population of tamarins on the island has decreased over the last 40 years. It might disappear completely if there is no further interference by man. Thus, the results of Hladik et al. are not so much definitive as suggestive and confirming.

Robinson (1973) describes some of the methods by which *geoffroyi* individuals discover "hidden" (cryptic or mimetic) insects. The insects may be baffling, but the tamarins can learn to cope with them. There are other accounts of hunting and prey-catching in Muckenhirn (1967), and Eisenberg and Leyhausen (1972).

S. fuscicollis occurs in the west of the Amazon basin from southern Colombia to southern Peru (Hershkovitz, 1966). It is divided into a number of slightly different looking populations, several of which have been recognized formally as subspecies. I was able to observe some of these animals in different parts of Colombia, in the *intendencia* of the Putumayo on the same island in the Rio Guineo as *Callimico*, and around the localities called El Pepino and Rumiyaco near Mocoa at approximately 700 meters above sea level, and even more frequently in the *intendencia* of the Caquetá, in the environs of the town of Valparaíso at somewhat lower altitudes. According to the latest systematic revision (Hershkovitz, 1968), all the individuals observed should have been representatives of the subspecies *fuscus*.

The climate of the Caquetá recalls the Putumayo and the Atlantic rather than the Pacific coast of Panama. It is quite humid. The apparent climax vegetation of most of the places that I visited in both *intendencias*, and all the well-drained sites, is tall and comparatively spacious forest. The resident *fuscicollis* resemble *geoffroyi* in preferring dense tangles and crowded clusters of trees of low to medium height. In the circumstances, they are (have to be) rather more catholic than their Panamanian relatives. Although they are common in many of the areas of second growth produced by man, they also inhabit the second growth of windfalls within the taller forest, and range into or along the borders of the thicketlike vegetation, which is all that can grow in certain waterlogged or swampy places. Both windfalls and swamps are rather numerous, but probably were scattered in the Putumayo and Caquetá under natural conditions. It seems likely, therefore, that the distribution of *fuscicollis* always has been patchier than was that of *geoffroyi* at one time. This may help to explain why *fuscicollis* is absent from some areas, in the Putumayo and at least one other part of Colombia, which appear to be physically suitable for the species. Filtering through patches must often be slow. Perhaps *fuscicollis* individuals were not able to reach some areas before their potential niche was preoccupied by competitors (see also below, the discussion of *Callicebus moloch*).

On the island in the Rio Guineo, the *fuscicollis* move through many or all of the same trees as *Callimico* and raid the same plantations. The coexistence of the two species in this area may be facilitated by the difference in their sizes. The larger form probably tends to take larger fruits and arthropods on the average and over the long run.

Other species of *Saguinus* also live in dense and moderately low forest and/or scrub and tangles.

Hernández-Camacho and Cooper, op. cit., suggest that two forms of northern and central Colombia, *oedipus* and *leucopus*, have much the same habits as *geoffroyi*, to which they are obviously closely related. 1947541

I saw groups of another form, a relative of *fuscicollis*, in second growth patches within tall forest on well-drained ground, and in very mixed vegetation on semiflooded ground, on the left bank of the Rio Guamés, near the Kofan Indian settlement of Santa Rosa, again about 300 meters above sea level, in the *intendencia* of the Putumayo south of the Putumayo River itself. No specimens were collected, but the appearance of the animals conformed to the description and diagnosis of *S. graellsi* as given by Hershkovitz (1966). (This is perhaps surprising, as *graellsi* had not previously been recorded frequently from Colombia.) These presumed *graellsi* individuals occur in precisely the sites where one would expect to find *fuscicollis* individuals, were any of the latter present in the area. (The two species are supposed to be sympatric in some places, but I could find no traces of *fuscicollis* around Santa Rosa, and the local Indians did not seem to be familiar with it.)

A third species related to both *fuscicollis* and *graellsi* and also Amazonian in distribution is *S. nigricollis*. It may be slightly more adaptable than some of the other tamarins. Hernández-Camacho and Cooper say that, in Colombia, it occupies a wide range of rain forest habitats from primary to secondary growth, including both seasonally flooded and nonflooded areas, and seems to show a preference for middle canopy levels.

S. inustus, a species of more distinctive appearance, has adapted to gallery forest and patches of woods in the savannah region of the valley of the Guaviare River, an affluent of the Orinoco (Le Nestour, personal communication, confirmed by Hernández-Camacho and Cooper).

Another form, which is not particularly closely related to any of the ones cited above, *S. midas*, is known to feed or forage in secondary forest and edge in the Amapá (Brazilian guiana), although it may also need taller and shadier forest into which to retire for protection from solar radiation in the middle of the day (Thorington, 1968a).

S. geoffroyi is primarily a springer, usually moving horizontally or diagonally. *S. fuscicollis fuscus* is primarily a vertical clinger and leaper, even more so than *Callimico* in my experience. *S. graellsi*, despite its other resemblances to *fuscicollis*, appears to be more like *geoffroyi* in its preferred method of locomotion. All three species of *Saguinus* usually range between 3 and 20 meters above ground. They probably average about the same as the *Callimico* of second growth, but lower than those of mature humid forest. They also come down to the ground from time to time. *S. geoffroyi* and *graellsi* seldom or never do so except when undisturbed, when there is no alternative and convenient route between trees during foraging or the search for social companions. *S. fuscicollis*, at least in the Caquetá, differs from its congeners and again resembles *Callimico* in coming down to the ground not infrequently in order to escape from potential predators.

The freedom of movement of both *geoffroyi* and *fuscicollis* in trees and bushes above ground is restricted in several ways, not only by whatever preference they may have for the vicinity of edge, tangles of vines, and relatively small trunks and branches, but also by a definite tendency to avoid certain types of plants, the large monocotyledons, such as palms. Probably their claws slip or slide on palm fronds, which are usually tough and smooth. The result of these restrictions is that the tamarins occupy or "cover" their territories less completely, presumably less intensively, than do many other monkeys.

The reactions of *geoffroyi* in the wild suggest that they must guard against numerous and varied predators. They are wary of birds of prey (except the Double-toothed Kite; see Chapter 4). They also react to dangers approaching on the ground. *S. fuscicollis* may have even more, or more serious, nonflying predators in some areas. Individuals of both species perform elaborate "mobbing" displays, with many conspicuous vocalizations, to distract mammalian carnivores before actual escape, but the performances of the Caquetá population of *fuscicollis* are often peculiarly vigorous and long sustained. (It would be useful to know how much the forms of antipredator reactions are affected by individual experience and group tradition. There are surprising differences among certain tamarins. Thus, for instance, the individuals of the population of *fuscicollis* near El Pepino and Rumiyaco are much less likely to perform mobbing and more likely to remain hidden and quiet than are the Caquetá animals. Their behavior could be a response to hunting by humans. Tamarins are shot for food in the Putumayo but not in the Caquetá. It may be significant, however, that the *graellsi* along the Guamués, which probably are hunted as enthusiastically and efficiently as their *fuscicollis* neighbors to the north, are considerably less prudent and tend to perform mobbing like *geoffroyi*.)

Apart from the Night Monkey, *Aotus*, all living species of ceboids are diurnal. Both *Saguinus geoffroyi* and *fuscicollis*, and probably many other tamarins, usually or often sleep in holes in trees at night. This may help them to keep warm, and should provide further protection against some predators, especially owls (see Chapter 6). G. A. Dawson (personal communication) has also found groups of *geoffroyi* sleeping huddled together in the open at the ends of branches of large trees. Possibly these individuals did not have suitable holes within their ranges or territories. Dawson was struck by the fact that the huddles looked like large termite nests at a distance, an apparent example of crypsis or protective mimicry.

Different species may have different activity rhythms during the day. *Saguinus geoffroyi* individuals often start to move rather late in the morning, well after dawn and appreciably later than most diurnal birds and some other monkeys. My impression is that *fuscicollis* and *graellsi* individuals may leave their sleeping quarters more promptly in favorable circumstances. All three species remain active throughout the middle of the day in fine weather. *S. geoffroyi* and *fuscicollis* are difficult or impossible to find during heavy rains at any time of the day. I think that many of them must remain in or return to their sleeping holes when the rain is heavy or prolonged. (This certainly is true of *geoffroyi* in captivity.) Individuals of *graellsi*, on the other hand, are sometimes active during the heaviest downpours, or take only partial shelter under overhanging leaves. This could be interpreted as a special adaptation to particularly humid environments, one of the factors that permits the species to coexist or compete against the flexible and opportunistic *fuscicollis*, which has a more widespread distribution.

Marmosets (*Callithrix* and *Cebuella*)

Marmosets share many characteristics with tamarins—claws and some locomotory and feeding habits, as well as other correlates of small size—but it is possible that the two subgroups are not more closely related to one another phylogenetically than either is to some other ceboids such as *Aotus* or *Callicebus*. (The complexities of the relevant behavioral evidence are noted in Moynihan, 1970b.)

Most marmosets are more soberly colored than the flamboyant kinds of tamarins. They tend to be grayish or buffy, with patches of black or white. They also have erectile tufts and ruffs around the face and head.

Two genera are usually recognized: *Cebuella* and *Callithrix*. They are very similar in basic anatomy. It might be reasonable to lump them together in a single genus; but they have different distributions and would appear to differ appreciably in a few behavior patterns, so it may be convenient to retain the generic distinction, at least for the time being and for the purpose of this account.

Cebuella seems to include only one species, the Pygmy Marmoset, *pygmaea*. It is very small indeed, by far the smallest of living monkeys. It probably is restricted to the upper part of the Amazon basin. There are descriptions of some aspects of its behavior and ecology in the Putumayo of Colombia in Moynihan (in press) and Hernández-Camacho and Cooper (op. cit.), and along the Nanay River southwest of Iquitos, Peru, in Ramirez et al. (in press).

The original habitat preferences of the species must remain something of a matter for speculation in view of past and current changes in the environments of the areas in which it has been studied. Ramirez et al. found it in dense vegetation below the canopy of "inundatable" and presumably old forest. R. Mittermeier (personal communication) saw it at the edge of a forest that was actually flooded (again near Iquitos). Hernández-Camacho and Cooper say that it is typically an inhabitant of mature nonflooded forest, and usually occurs in association with "guarango" trees (*Parkia* sp.), often emergent over the canopy. H. Le Nestour (personal communication), and some of the local Indians with whom I talked in the Putumayo, saw individuals and groups along a variety of forest edges. My own observations were made in and around El Pepino and Rumiyaco. Here the Pygmy Marmosets seem to be most abundant in "hedges," small and often isolated strips and patches of relatively low and degraded or unnatural woodland, which the local settlers have left or planted between pastures and cropfields.

The usual locomotory movements of individuals of the species when undisturbed are much like those of *Saguinus fuscicollis*. Mostly vertical clinging (see Figures 6 and 7), climbing, and leaping, with some interspersed quadrupedal and more or less horizontal running and springing. These marmosets may be even more vertically oriented than any of the tamarins, simply because they spend a great deal of time moving up and down tree trunks while feeding (see below). They also come to the ground to look for food or to move from one food source to another.

The spectrum of foods taken is obviously broad, tamarinlike in some respects, peculiar in another. Pygmy Marmosets eat insects, spiders, and berries and probably some buds in trees

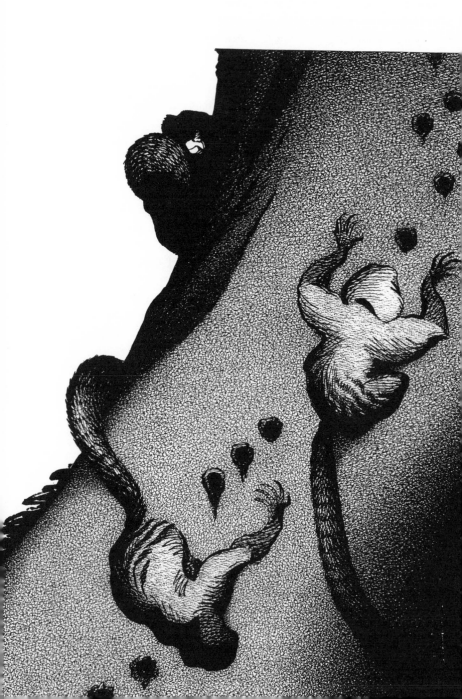

Figure 6. Pygmy Marmosets on a feeding tree.

and bushes. They come to traps baited with exotic, non-native fruits such as plantains and bananas. They have been seen to descend to cleared pastureland to catch grasshoppers.

Much more distinctive is "sap-sucking." Every family or group seems to own one or more large or thick trees, the trunks and larger branches of which are riddled with small and shallow holes (Figure 6). These holes would appear to be drilled, presumably gnawed, by the marmosets themselves. They have specialized procumbent lower incisors, which may help in the process. The holes are visited very frequently and repeatedly by their owners, who nuzzle into them at each visit, evidently extracting something of interest or value. More often than not, this must be sap (or possibly some other kind of plant "gum"). In fact, Hernández-Camacho and Cooper found sap in the stomachs of specimens collected at Puerto Leguízamo, not far from El Pepino and Rumiyaco. The visits of the marmosets to their feeding holes are frequent enough to suggest that sap may be the most important of all the elements in their diet.

According to Hernández-Camacho and Cooper, *Parkia* is the usual source of sap under natural conditions in Colombia. During my own observations of hedge animals, I found feeding holes in several different species, including an *Inga* (probably *spectabilis*—supposedly introduced into the Putumayo), *Matisia cordata*, and something called "*cedro*" by the local settlers, possibly *Cedrela odorata*. The preferred sap trees of the Peruvian animals observed by Ramirez et al. were *Quararibea* sp. (Bombacaceae). Other species such as *Policourea macrobotrys* were also attacked occasionally.

The marmosets of El Pepino and Rumiyaco, using hedges and exotic foods, have certainly become commensals of man. All or most individuals of the species are active throughout most of the day in fine weather, but may rest around noon.

They are extremely wary of potential predators. They have a whole series of antipredator devices, perhaps more than any other New World primate. All these devices seem to be designed to avoid attracting the attention of a predator rather than distracting him. Pygmy Marmosets do not perform loud mobbing like tamarins and other monkeys. In the field, they are the quietest of ceboids (at least in the range of frequencies audible to human ears). They also try to keep out of sight. They dodge behind trunks and branches like squirrels. They have developed special protective types of locomotion. Sometimes they move very, very slowly, barely oozing forward or sideways (Figure 7). As in the case of sloths, this makes the movements inconspicuous. More often, they advance in spurts, lizardlike alternations of dashes or leaps and frozen immobility. Their coloration, largely buffy, grizzled or brindled with blackish, is highly cryptic. They are incredibly difficult to detect when motionless. Perhaps their most feared predators are birds of prey. They make special efforts to keep away from the canopy and, thus, to minimize exposure to anything flying overhead. (Birds of prey tend to become rare in the immediate vicinity of human settlements. This must be another advantage of such areas from the point of view of the marmosets.) The *Cebuella* that I observed slept in holes in trees, at some distance from their feeding trees.

Marmosets of the genus *Callithrix* are larger than Pygmies. They occur in southern and eastern Brazil, extending to the south bank of the Amazon in some places. There are several species, plus some very well marked subspecies or semispecies, but perhaps no more than two species groups (see Appendix 2). A typical form, *Callithrix geoffroyi* (or *C. [jacchus] geoffroyi*) is shown in Figure 8.

We do not have much information on the behavior and ecology of these marmosets in their native habitats. Such data as are available (Cabrera and Yepes, 1940, Cruz Lima, 1945, Hill, 1957, Coimbra-Filho, 1971, and Coimbra-Filho and Mittermeier, 1973a) suggest that some or all of them fill much the same ecological niche(s) as tamarins. One would like to know how competition between tamarins and marmosets is con-

Figure 7. Hostile displays of Pygmy Marmosets. The top individual shows extreme piloerection (a "General Ruffle" with exaggerated "Tail-ruffling"). The lower individual illustrates one of the few ritualized facial expressions of the species, with spreading of the hairs on both sides of the head and a peculiar posture sometimes assumed during the slow "Oozing" escape.

Figure 8. Callithrix (jacchus) geoffroyi. *The two top individuals show varying degrees of erection of the hairs around the face and ears, apparently in hostile display.*

trolled or avoided, or when it is not. Probably size is relevant here too. The two principal genera of medium size for the tamarin-marmoset assemblage as a whole, *Saguinus* and *Callithrix*, apparently do not overlap in geographical range. They may be actively mutually exclusive. The only marmoset that is known to overlap with *Saguinus* is the much smaller *Cebuella*. The only tamarins that overlap broadly with *Callithrix* are forms of *Leontopithecus*. This last case must involve some additional complications. Coimbra-Filho and Mittermeier note that *Callithrix* (*jacchus*) *penicillata* is more tolerant and adaptable than *Leontopithecus*. It extends into many sorts of second growth and plantations as well as "primary" forest. The peculiar hands and fingers of *Leontopithecus* and their possible use in gathering food have already been mentioned. Some or all of the forms of *Callithrix* also share the sap-sucking proclivities of *Cebuella*. They have similar procumbent lower incisors. Some *geoffroyi* marmosets that I watched in the National Zoo in Washington chewed on dead branches in their cage with vigor and persistence. More nearly conclusively, individuals of *penicillata* have been seen, by Coimbra-Filho, to perforate the bark of trees in the wild. Doubtless further studies will extend and refine our understanding of these distinctions.

The Night Monkey (*Aotus*)

This is one of the more widely distributed ceboids, ranging from northern Argentina and Paraguay to western Panama (and possibly farther north and west in Central America). There are appreciable morphological differences among populations of Night Monkeys of different areas and among different individuals of the same areas, but Hershkovitz (1949) includes them all in the single species *trivirgatus*. (Variations in serum proteins and chromosome structures and numbers are also known—see Brumback and Willenborg, 1973, and Benirschke et al., 1974. They are difficult to interpret or to assess for purposes of classification, which raises a general problem.

The available information on the cytogenetics of New World primates, summarized in De Boer, 1974, is remarkable. There are variations within species of several genera, and enormous differences between species that are certainly closely related to one another. At the moment, the information is more interesting than useful.)

Night Monkeys are somewhat larger than the largest tamarins and marmosets. They are quite similar to the titi monkeys of the genus *Callicebus* in size and general body proportions, but they have much larger eyes. The superficial similarity between *Aotus* and *Callicebus* is so great that the two were often associated with one another in earlier classifications. Some of the resemblances must be significant, but they are not necessarily indicative of intimate phylogenetic connection. They may be primitive and/or correlated with size and/or adaptations to certain feeding habits and methods of locomotion. Both titis and Night Monkeys are "springers," having more efficient grasping hands and feet than tamarins or marmosets. (At least, they really do grasp a greater variety of branches more frequently during locomotion. Their nails are not particularly clawlike. They usually apply to the fleshy pads of the fingers and toes, rather than the terminal nails, to obtain purchase on a substrate. Apart from this, their feet do not differ very much from those of tamarins and marmosets, and those of the remaining species considered below do not differ very much from those of *Aotus*. Feet would appear to have been conservative during the evolution of the ceboids. The fingers of the hands of Night Monkeys and titis, although radially arranged, may be capable of more independent or flexible movement than those of *Saguinus* or *Cebuella*. The distinction is slight at best, however. Neither titis nor Night Monkeys have developed manipulative habits to any great extent.) The known characters of Night Monkeys, which are not obviously immediately adaptive, such as some aspects of their vocalizations, would

Figure 9. A high intensity aggressive display, the "Arch Posture," of a Night Monkey.

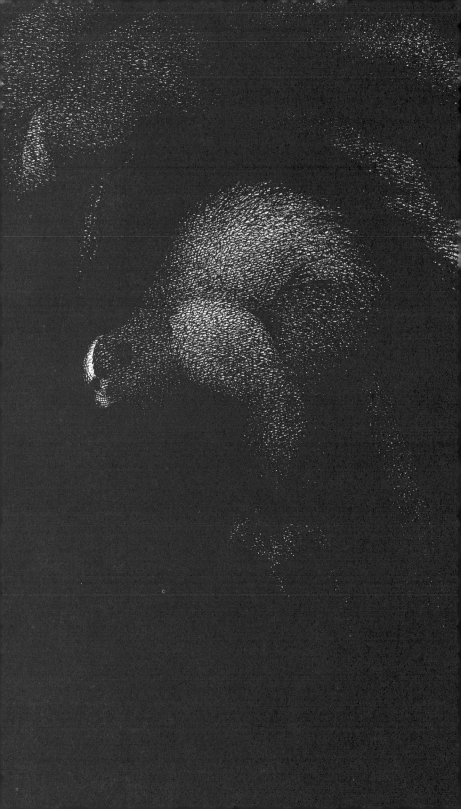

suggest that they may be more closely related to tamarins than to anything else, or to both tamarins and howler monkeys.

The fur of Night Monkeys is soft, dense, and comparatively short. Different individuals may be various shades of brown or gray above. They usually are buffy or caramel-colored below. The distal part of the tail is black. There is a rather intricate arrangement of black and white stripes and spots on the face, forehead, and crown. See Figure 9.

This kind of pelage must be advantageous. It has been evolved by many other nocturnal arboreal mammals, including the lorises of southern Asia, some of the phalangerids of Australia and New Guinea, and some of the opossums of tropical America. Even the plumages of some owls are not very different. The coloration of the body seems to be cryptic on the whole, while the dark and light of the head can be disruptive when viewed by potential predators, or a useful adjunct to visual signaling during intraspecific encounters.

The species has been studied most intensively in central Panama (Moynihan, 1964, Thorington et al., in press). It occurs in many different forests, monsoon and rain forests (see Figures 10, 11, 12, and 13), as well as moderately old and tall second growth. In the latter, it may be almost as much a regular commensal of man as *Saguinus geoffroyi*, although probably less so than some *Cebuella pygmaea*.

Tokuda (1969) suggests that the Night Monkeys of the Putumayo prefer the lower strata of forest. Panamanian individuals range through many levels of vegetation. They are usually found between 7 and 25 meters above ground, although they may go higher or lower on occasion. They will enter almost all kinds of trees, including palms and other large monocotyledons. They have not been seen to come down to the ground under natural conditions; in captivity they do so quite frequently. They move easily on flat or broad surfaces. This would suggest that they are capable of coming down to the ground in the wild whenever necessary or desirable in emergencies. If so, this might help to explain why the species has spread so widely.

Precise figures on population numbers of the smaller ceboids are difficult to come by. All I could say about tamarins and marmosets is that such forms as *Saguinus geoffroyi, S. fusci-collis*, and *Cebuella pygmaea* are at least fairly abundant in some areas. For *Aotus*, N. Durham (personal communication) hazards a slightly more quantitative estimate, 25 to 50 individuals per square kilometer, in lowland forests of south-eastern Peru.

Night Monkeys eat a variety of foods, fruits, buds, and insects. In Panama, they appear to take relatively more fruits and fewer insects than do tamarins. The same may be true of individuals far to the south, in Peru (Durham) and in the province of Santa Cruz in Bolivia near the Paraguayan border (Krieg, 1930). Yet the species probably is not as nearly completely vegetarian as *Callicebus moloch* or the more familiar forms of *Alouatta*. See also C. M. Hladik et al. (op. cit.).

Very little is known about predation upon Night Monkeys. Only some indirect evidence may be revealing. Panamanian individuals are often rather noisy when they are active in the wild. They tend to utter many "contact" or locomotory notes, which may indicate that they are not particularly vulnerable to predators as long as they continue to move about. They usually or always sleep in holes in trees during the daytime. They do not seem to occupy the same holes as tamarins.

Howler Monkeys (*Alouatta*)

These animals are quite different in outward appearance from any of those described above. They look much more typically monkeylike. They are large, among the heaviest, if not the tallest, of New World primates. They also have prehensile tails.

There is some evidence, nevertheless, that howlers may be more closely related to the Night Monkey than to most other

living ceboids. They have a few possibly significantly similar behavior patterns and morphological characters (see also Stirton, op. cit.).

The genus *Alouatta* is even more widely distributed than *Aotus*, ranging from the northern border of the humid tropics in Central America to the southern border in Bolivia, Paraguay, and the northernmost parts of Argentina. Many forms can be distinguished. They are largely allopatric, nonoverlapping, but there are one or two areas where two forms have been reported to come together. Some populations also differ from one another in various features in ways that would be expected to discourage interbreeding (see below). It would seem, therefore, that several species must be recognized.

The one that has been studied most thoroughly has been called both *villosa* and *palliata* in the literature. I shall use the former name here, without prejudice (but see also comments in

Figure 10. A close-up view of the middle level of a humid "monsoon" forest in the Atlantic drainage of Panama (Barro Colorado Island). Photo by O. Linares.

Figure 11. A wider view of the middle level of forest on Barro Colorado Island. Note the wide spacing between trees, clumps of leaves, and branches. Photo by O. Linares.

Figure 12. Edge of monsoon forest in the Canal Zone Forest Reserve in Panama. Photo by O. Linares.

Figure 13. Canopy of monsoon forest in the Canal Zone Forest Reserve. Photo by O. Linares.

J. D. Smith, 1970). This species ranges from Mexico into northern Colombia. Some of the central Panamanian populations, especially the animals on Barro Colorado Island, were the subjects of a famous pioneering investigation by Carpenter (1934). He discussed them again in later years (e.g., 1962 and 1965) and his observations have been continued and extended in some directions by other students at irregular intervals (Collias and Southwick, 1952, Altmann, 1959, Bernstein, 1964, Moynihan, 1967, A. Hladik and C. M. Hladik, 1969, Chivers, 1969, C. C. Smith, personal communication, Richard, 1970, C. M. Hladik et al., op. cit., Mittermeier, 1973, Thorington, personal communication). Other populations of the same species have been studied in Costa Rica (Heltne et al., in press, Freese, in press) and in Chiriquí in southwestern Panama (Baldwin and Baldwin, 1972a, and in press). Individuals of a similar form, *pigra*, of uncertain status (see Appendix 2), have been studied at Tikal in the Petén region of Guatemala (Coelho et al., 1974).

There also have been observations of two rather different species: *seniculus* in Venezuela and Trinidad (Neville, 1972a and in press), and in Colombia (Klein and Klein, in press, and Hernández-Camacho and Cooper, op. cit.—I saw a few individuals myself in the Caquetá, and learned something of the supposed status of the species in the Putumayo); and *caraya* on islands in the river Paraná in the province of Corrientes, northern Argentina (Locker Pope, 1968).

All the forms have comparatively simple color patterns. Juvenile and adult, male and female, *villosa* individuals are largely blackish (Figures 14, 15, and 16). So are typical *pigra*. All individuals of *seniculus* are more or less brilliant orange-rufous or bronze color. *A. caraya* differs from the other species in being conspicuously sexually dimorphic in color; adult males are black but adult females are brown. In all forms, the pelage tends to be silky. It usually is ornamented by some simple nonerectile tufts or fringes of long hair, most notably beards. These are much more highly developed in adult males

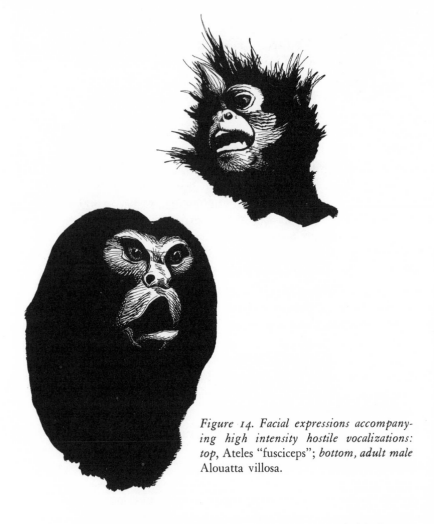

Figure 14. Facial expressions accompanying high intensity hostile vocalizations: top, Ateles "fusciceps"; bottom, adult male Alouatta villosa.

than in females or juveniles. An even more remarkable feature of the genus is a specialization of the hyoid apparatus, the resonating chamber for the voice. Several elements of the structure have become greatly enlarged and ossified, facilitating the production of loud and long noises. Howlers are justly famous for the volume of the sounds that they can, and frequently do, utter. It is this talent that has earned them their vernacular name (even though the actual sounds are more

*Figure 15. Relaxed postures of adult male
Alouatta villosa.*

Figure 16. Infants. On the left, a young lagotricha. *On the right, a young* Alouatta villosa. *Both show purely or partly hostile expressions.*

often roars or barks than howls). The specialization of the hyoid is most exaggerated in adult males, and it also differs from species to species, being most extreme in *seniculus* and least in *villosa*. It is suggestive, therefore, that the beards of *seniculus* males (Figure 17) are considerably longer than those of *villosa* males. The beard covers the throat and draws attention to that part of the body. The positive correlation between the elaboration of the hyoid, to emphasize sounds, and the lengthening of the beard, which can only be seen (or felt), would seem to be an interesting example of summation of stimuli, in effect social signals, mediated by different senses.

Howlers are highly arboreal and adapted to life in large trees. They need fairly large branches and space to move around in simply because of their size and weight. One can hardly imagine them remaining comfortably or safely for any length of time in dense thickets or tangles of frail twigs and vines. They are basically quadrupedal, walking, pacing, running, and leaping along branches. In some circumstances, they show occasional or incipient tendencies to swing by their arms, but they do so less frequently than spider monkeys or the Woolly Monkey. They make much use of their prehensile tails. When moving quadrupedally, they usually keep the tail pressed against or curled around a branch as an extra support (if only for moral reassurance). They can also hang by the tail while feeding or playing, but seldom by the tail alone. More often than not, they continue to hold on with one hand or foot at the same time.

Prehensility of the tail has been evolved by several other kinds of large ceboids, possibly independently, and it is also characteristic of many other tropical American mammals of other groups (opossums, anteaters, a porcupine, *Coendou*, and the Kinkajou, *Potos*, one of the Carnivora). The repeated development of this specialization among diverse animals must be another indirect reflection of the predominance of forest habitats in the New World tropics.

The hands of howlers are greatly modified in a peculiar way for grasping of a limited type. The first two fingers of each hand are set apart from, and are opposable to, the other three. This facilitates the firm clasping of branches but impedes more delicate forms of manipulation. (The arboreal chameleons have evolved a very similar mechanism.) As would be expected, therefore, howlers handle detached objects relatively rarely, probably even less frequently than do tamarins, marmosets, or Night Monkeys. Individuals of *villosa* perform comparatively little grooming, less self-grooming than any other

New World primate with which I am familiar, and less mutual grooming than any other type except *Saimiri*. Possibly as a consequence, they suffer severely from external parasites, especially botfly larvae. (It may also be noted that the tail is purely a supporting and locomotory organ among howlers. In spite of its inherent flexibility, it is used to seize independent objects even less frequently than are the hands.)

Howlers can move along the ground quite effectively in captivity but do not seem to like it very much. In the wild, *villosa* individuals come down to the ground only when absolutely "forced" to do so (when attacked by another species, or as a result of a fall, or to rescue a fallen infant, or when the forest itself is cleared). Individual *seniculus* may be somewhat less reluctant. The human residents of the Caquetá claim to have seen howlers come down to drink from springs and small marshes near the heads of streams in the forest. As the waters of at least some of these springs and streams are highly mineralized, the animals may be getting some useful or necessary nutrient from them (see also below). According to Hernández-Camacho and Cooper, *seniculus* have been seen crossing large expanses of savannah in some areas, presumably passing from one forest to another, and they may also swim across large bodies of water 200 to 300 meters wide.

Leaving aside excursions, it seems probable that the preferred habitats of all or most howlers are tall and more or less old, mature or maturing, humid forests on usually well drained ground, just the sorts of habitats in which they occur on Barro Colorado and in parts of the Caquetá now. But they have a wide, if perhaps temporary, tolerance of other kinds of forests. They inhabit both dry riparian forest in the province of Guanacaste in Costa Rica (Heltne et al., op. cit., Freese, op. cit.) and coastal swamp forest in Chiriquí (Baldwin and Baldwin, in press).

Figure 17. Red beards and sideburns: top left, Pithecia "rubicunda"; *top right, adult male* Alouatta seniculus; *bottom*, Pithecia "rubicunda." *All three individuals are relaxed.*

Wherever they occur in trees, they tend to range from only a few (5 to 8) meters above ground up to the canopy.

There is a good deal of information on the foods and feeding habits of some of the forms. Hernández-Camacho and Cooper say that the *seniculus* of Colombia take leaves and some soft fruits of *Anacardium excelsum*, *Cecropia* spp., *Cassia moschata*, *Ficus* spp., *Spondias monbin*, and other forest plants. They may also tap other resources of a different kind (in addition to mineral water). Local woodcutters in the Caquetá claim that individuals of this species gnaw or chew on dead and dying tree trunks and branches, supposedly in order to extract termites. I have seen trees that had been gnawed by something, where howlers had been seen a few days earlier by other people, but unfortunately the animals never reappeared while I was there. According to Coelho et al., the *pigra* of Tikal show a marked preference for the fruits of *Brosimum alicastrum*. The diet of the *villosa* of Barro Colorado Island has been studied by the Hladiks and their collaborators (works cited above). The species is also primarily vegetarian, taking both leaves and fruits. Eighty per cent of the ingested vegetable material is supposed to come from the following plants: *Anacardium excelsum*, *Spondias monbin*, *Ceiba pentendra*, *Quararibea asterolepis*, *Dipteryx panamensis*, *Inga* spp., *Platypodium maxonianum*, *Brosimum bernadettae*, *Cecropia* spp., *Ficus* spp., *Peribea xanthochyma*, and *Virola panamensis*. The figs, especially *Ficus insipida* (sic) are particularly important; *insipida* may comprise 50% of the material. *A. villosa* differs from the other fruit-eating ceboids of the area in taking much of its fruit in a very unripe state. It also gets some insect food in the form of grubs (species of *Blastophaga* for example) within fruits. These may be swallowed "accidentally" rather than deliberately sought, but they must provide a significant supplement to the rest of the diet. The Hladiks et al. calculate the contents of the foods of the Barro Colorado howlers as 9.6% protids, 3.2% lipids, 21.7% reducing glucids, 13.6% cellulose, and 51.9% complementary fractions.

Alouatta spp. are the only American monkeys known to feed on leaves to a very appreciable extent. Interestingly enough, *caraya* (Cramer, 1968) and *villosa*, and presumably the other species of the genus, differ from the leaf-eating monkeys of the Old World, *Colobus* and related forms, in having stomachs of a rather simple construction or arrangement. K. Milton (personal communication) suggests that howlers concentrate on the newest and tenderest leaves. This would simplify digestion. Fermentation of the more difficult materials may also take place in the caecum or colon.

Some of the foods of *villosa* are widely distributed and abundant throughout the year on Barro Colorado and in many other parts of Panama under present conditions. This may be correlated with the characteristic "tempo" of the species. *A. villosa* individuals are sluggish. Most groups of the species have medium to large territories. They cover them eventually, but they tend to do so slowly and intermittently. Presumably they never have to engage in active pursuit of, or prolonged search for, food. Individuals of the species certainly are much less frequently active than all or most other ceboids (*Callicebus moloch* may be as "tied down," but in quite a different way— see below). At times, *villosa* are slothlike. Sloths do, in fact, eat some of the same foods.

Howlers sleep on branches or in the crotches of trees, not in holes. Obviously they could not find enough holes of adequately large dimensions. They also seem to be less afraid of predators than many of their smaller relatives. They react to mammalian predators, but often ignore flying birds of prey.

These features may help to explain another. Howler monkeys have relatively smaller and apparently simpler brains than any other large ceboids. Perhaps they are confronted with fewer problems, or usually have more time in which to solve their problems. See the discussion in Chapter 7.

Everything considered, *Alouatta* would seem to be one of the most peculiar of New World primates. Only *Cebuella*, *Ateles*, and *Cebus* itself are equally exaggerated.

A further point, perhaps a qualification, may be mentioned. Populations of howlers are different in different areas. There also are average differences in densities of different species, some of which may be intrinsic to the species, others of which probably are not.

All the known *villosa* populations seem to be "artificial," although not always in the same ways. The howlers of Guanacaste are rare, scattered, reduced by the clearing of forests, the widespread use of pesticides, etc., and are generally in a very precarious state. Other populations of the species are larger, but certainly not stable. For a long but unspecified period of time, the howlers of Barro Colorado Island suffered serious epidemics of yellow fever at roughly seven-year intervals, with crashes in the bad years and booms in between. The situation has changed recently. The last two expected epidemics never occurred. (The reason is fairly obvious. The epidemics used to come from the Darien in eastern Panama. Much of the forest between the Darien and Barro Colorado has been cut down as the countryside has been settled and developed. This has interrupted the movements of the insect vectors of the disease.) It is difficult to interpret some of the results. Yellow fever itself is supposed to have been introduced to the New World some time after A.D. 1492. And Barro Colorado is slightly peculiar: it does not support many large mammalian predators, certainly no Jaguars (*Panthera onca*) for many years, and perhaps few of the other large cats. More important, as noted above, it is monsoon forest just approaching maturity. My impression is that submature forest of this type is richer in fig trees than is completely mature forest at the end of succession. As a result of some or all of these factors, the local population of howlers has been increasing rapidly (Chivers, op. cit., and Eisenberg and Thorington, op. cit.). Populations of *villosa* appear to be even larger in Chiriquí. Baldwin and Baldwin (in press) state that the density of howlers in the forests that they visited, during the period of their observations, was 104 individuals per 10 hectares (25 acres). This is enormous, approximately 12 times Chivers' last estimate for

Barro Colorado. Perhaps the Baldwins' populations were swelled by refugees from other areas destroyed in recent years. Possibly tree sloths were less common in the Baldwins' forests than on Barro Colorado (Montgomery and Sunquist, in press). The howlers of Chiriquí may have little competition.

A. caraya would appear to be at least moderately abundant on its islands in the Paraná. These islands also lack other species that occur on the mainland (Malinow, 1968), presumably because they have been shot out.

The reported high densities of some howlers are without parallel in the areas of the Caquetá and the Putumayo with which I am familiar. Wherever the tall forest survives in Amazonian Colombia, the fauna remains more nearly intact than farther north or in the far south. Fig trees are less conspicuously dominant than on Barro Colorado. Perhaps significantly, the local *seniculus* seem to differ from both *villosa* and *caraya* in being usually or always very scarce, in my experience the rarest of resident monkeys. They may also be less sluggish than the *villosa* of Panama. This is the more remarkable because they are not hunted in the Caquetá and probably only infrequently in the Putumayo. Possibly I found *seniculus* at the nadir of a population cycle. Yet the local settlers are unanimous in asserting that the species is never very numerous, and this seems to be confirmed by Tokuda's observations (op. cit.) at other (mostly down river) sites in the Putumayo. Klein and Klein (op. cit.) also found *seniculus* to be relatively scarce in the National Park of La Macarena in the Meta region of Colombia. The population censused by Neville in Venezuela was larger but far from crowded in terms of individuals per unit area. Walsh and Gannon (1967) rescued many howlers of the same species from a flooded valley in Surinam. Their data, such as they are, suggest that the local population was flourishing before the catastrophe, although it was still inferior to that of Barro Colorado.

It is at least conceivable that howlers reach their maximum potential only in abnormal conditions or peripheral regions.

Sakis and Uakaris (*Pithecia*—including *"Chiropotes"* and *"Cacajao"*)

This assemblage is diverse. It includes a number of forms. Several are allopatric; others are sympatric. They differ among themselves in bodily proportions as well as pelage and other superficial characters. There are, however, partly intermediate types between extremes. The various forms certainly must be assigned to different species, perhaps four superspecies, but they may all be lumped together in a single genus.

The distribution of the genus is essentially Amazonian, with extensions to the Orinoco valley and the guianas, but not to Central America or southeastern Brazil. Some of the individual species are supposed to have very restricted ranges within this general distribution.

The combination of appreciable speciation with considerable morphological diversity among species would suggest that the assemblage is comparatively old. (If Stirton's interpretation of one of the fossil types, *Cebupithecia*, were eventually to be confirmed, it would prove that the stock was already distinct from other ceboids by the Miocene. This would seem to be probable in any case.)

Sakis seem to be rather less specialized than uakaris. They show points of resemblance to howlers in details of anatomy and appearance. At least one form recalls tamarins of the genus *Saguinus* in some aspects of infantile behavior.

They come in two sizes: medium and medium large. The former is the genus *Pithecia sensu stricto* of older workers. There are supposed to be two species of this kind: *pithecia* and *monacha*. Possibly they will eventually be found to be only subspecifically distinct. The maps and summaries of their ranges given by some authors, e.g. Hill (1960), which imply extensive overlapping, must be at least partly mistaken. Cruz Lima (op. cit.) suggests that they are strictly allopatric. Until the question is settled, the two names can be retained for convenience *pithecia* for animals of the Guiana region and *monacha* for the most similar animals of the upper Amazon.

They have long bodies, moderately long arms and legs, and relatively very long, heavy, and completely nonprehensile tails. Their pelage is distinctive. Most of the head (except the face), the back and sides, and the tail are covered by long, rather coarse, grizzled blackish hair. This greatly increases the apparent size of the animals. They look almost as massive as howlers or Woolly Monkeys in the field. The underparts are less densely furred. The central part of the face is dark, largely due to dark skin showing through a covering of short and sparse hairs. There are very considerable differences among different populations and age and sex classes. There also is a great deal of individual variation, more than in other ceboids. Most animals have rather conspicuous, tamarin-like, white stripes across the cheeks. Male *pithecia*, especially adults, have a conspicuous "mask" of cream-colored or bright yellow hairs of a rather feathery texture around the face. Adult male *monacha* and some adult females may show less conspicuous or specialized versions of an equivalent mask in gray. The gray varies enormously, from whitish to dark sooty. In a few individuals it is suffused with brown. The general effects are difficult to describe in words, but Figure 18 shows some dark-faced individuals of *monacha* and a male *pithecia*. Figure 19 shows some of the lighter-faced individuals of *monacha*.

Both of the supposed species have been kept in captivity, although not usually in optimal conditions or with great success. I have also seen many *monacha* in the Caquetá and a few in the Putumayo. Tokuda saw other individuals of the same form elsewhere in the Putumayo.

They occur in several kinds of vegetation. Tokuda found them in the second stratum of very tall forest, around 20 meters above ground. In the Caquetá, they occur in stands of old tall and humid forest and, perhaps more frequently, in mixed forest that includes many moderately tall trees, forming a semicontinuous or reticular canopy over shorter second growth, presumably the result of selective cutting by man. In these circumstances, they seem to prefer the tallest trees available, usually ranging from 10 to 35 meters above ground. They may pass through second growth scrub beneath trees, occasionally

Figure 18. Small sakis: top and center, hostile facial expressions of young Pithecia "monacha"; *bottom, arch Posture of young* P. pithecia s.s.

coming as low as 4 or 5 meters, but they apparently never remain there for any length of time. It is my impression that they range higher on the *average* than any of the species previously cited, with the possible exception of howlers. (I saw some *monacha* go higher than any *Alouatta seniculus* in the areas that I visited; but Tokuda suggests that these howlers are characteristic of higher levels of vegetation than are the sakis in the areas that he visited. The difference, if real, may be ascribed to variations in the behavior of *seniculus*.) At whatever levels they move, *monacha* individuals enter all types of trees; they do not hesitate about palms.

As might be expected, their methods of locomotion are somewhat flexible. They are primarily quadrupedal walkers and leapers. When undisturbed, they usually trundle along rather slowly and quietly. But they can gallop at astonishing speed and leap prodigious distances (Figure 19) in almost any direction whenever they want to—in escape, or play, or for any other reason. Their common name in the Caquetá and Putumayo is "volador"—flier, which is really very appropriate. Individuals may also swing by their arms from time to time. Sanderson (1957) reports upright walking by some form of saki, probably *pithecia*, in the wild. I never observed this in *monacha* under natural conditions, but both adults and young frequently stand up on their hind legs alone without walking. They are most apt to do this in display during hostile encounters. See also Stern (1971).

The hands of both *pithecia* and *monacha* are built like those of *Alouatta*. They have the same peculiar 2–3 split.

It would seem that the smaller sakis are largely frugivorous. A few *pithecia* collected in Surinam (Fooden, 1964) had little or nothing but fruit in their digestive tracts. I saw the *monacha* individuals in the Caquetá take a variety of fruits, mostly of moderate size and some with tough skins. (All sakis and uakaris have procumbent incisors as well as large canines. The combination could be particularly useful in coping with tough fruit.) Nevertheless, Sanderson claims to have seen *pithecia*

catching and eating bats. This record may be dubious, but I certainly have seen *monacha* individuals in the Caquetá inspecting and turning over leaves, apparently searching for insects. Unfortunately, they were too high in the trees for me to be able to determine if they were successful or not. Probably they are less nearly completely monophagous than howlers when given a choice.

They may be too large to sleep in holes.

They are quite wary of man and probably many other potential predators.

The larger sakis were once placed in a genus *"Chiropotes."* They comprise a few forms, possibly all belonging to a single species, distributed in parts of the lower Amazonian and guianese regions. They differ from *pithecia* and *monacha* in minor features of the skull and in having more complexly convoluted brains. They also are bearded. Otherwise, they are quite like the smaller forms in morphology. They have long hair on the body, usually black or black and brown in color, and long nonprehensile tails. Very little is known of their behavior or ecology in the wild. Possibly they are not very distinctive. The local form of larger saki which occurs in parts of Surinam seems to be frugivorous (Fooden, op. cit.) and to inhabit tall trees in humid forest (Husson, 1957).

The uakaris ("*Cacajao*") are equally large and show the same features of skull and brain. But the two most extreme forms, *calva* and *rubicunda*, are the most peculiar looking of New World primates. Both have bare faces and partly bald heads of a brilliant raspberry red color, and long silky hair on body, limbs, and tail. This is silvery white in *calva*, coppery orange in *rubicunda* (Figure 17). The effect is picturesque. Underneath the hair, the tails of all sakis are more or less abbreviated. Those of *calva* and *rubicunda* are less than half the length of the body.

The two forms are largely allopatric. Their different coloration may have a simple genetic basis. Mittermeier (personal

Figure 19. Adult Pithecia "monacha." *The top individual is in full flight (seen from below). The other two are almost relaxed.*

communication) has found whitish individuals in some *rubi-cunda* populations, as variants or as a result of introgression. Thus, the two forms probably should be placed in the same species, for which *calva* is the appropriate name. Following the precedent set for the smaller sakis, however, I shall continue to use both names singly without bothering to indicate trino-mials.

All uakaris are restricted to the upper Amazon. They have been observed only slightly more often than the larger sakis. The few published accounts are anecdotal. Perhaps the best is by Bates (1863). This account would indicate that *calva* is very sakilike, frugivorous, and an inhabitant of tall trees in high (in this case periodically flooded) forest. One may assume, nevertheless, that there are at least a few ecological differences between sakis and uakaris. There is even some reason to be-lieve that the differences could be considerable. *P. monacha*, and probably all other sakis, use their long and heavy tails as counterweights during rapid locomotion, especially when jump-ing. Obviously *calva* and *rubicunda* cannot do this. They prob-ably leap relatively less frequently, and run (on their hind legs?) and swing by their arms relatively more frequently. Sev-eral individual *rubicunda*, which were kept in captivity on Barro Colorado, also managed their hands in a way that was different from *monacha*. There was seldom any conspicuous indication of a 2–3 division of the fingers; nor were the thumbs strictly opposable. More often than not, all the fingers of a hand were kept together or spread radially as in *Aotus* and *Callicebus*.

Captive *rubicunda* in good health are among the most con-tinuously active, and alert, and apparently intelligent of all monkeys. Sakis of the *pithecia* and *monacha* type can be much more energetic and enterprising in captivity than might be supposed from some published accounts, but they are quite surpassed by *rubicunda*. A single captive young *calva* observed in the Bronx Zoo in New York was indistinguishable from *rubicunda* in behavior. Among the ceboids with which I am familiar, only the various species of *Cebus* itself convey the same impression of adaptability and insight under the highly aberrant conditions of cage life. This must be a reflection of some adaptation to more normal surroundings. Possibly uakaris have to do more complex things than all or most sakis in the wild.

The suggestion may be partly supported by data from another species, *melanocephala*. This form is black and brown, rather like some "*Chiropotes*," but it is usually bracketed with "*Cacajao*" and supposed to be a uakari. It has been observed by several students. Hernández-Camacho and Cooper say that it is an inhabitant of "high canopy." Le Nestour (personal communication) encountered a few groups of the species in apparently mature but only moderately high and not very rich forest in the Vaupés region of Colombia, downriver from the Caquetá and the Putumayo. Mittermeier (personal communication) found large populations in adjacent parts of Brazil, in "igapó" (seasonally flooded) forests along black water rivers and streams. He says that they feed on fruits, nuts, and insects. Le Nestour saw individuals eat leaves, eggs, berries, and tadpoles from small bodies of water in epiphytic plants, holes, and crannies of trunks and branches. It seems likely, therefore, that the species takes a greater variety of foods than do any of the typical sakis.

Titi Monkeys (*Callicebus*)

As already indicated, titi monkeys are moderately small and generalized in structure. They are undistinguished in external appearance except for their long and fluffy fur, which is brilliantly colored in some forms. Typical facial patterns are shown in Figure 20. Apart from similarities to *Aotus*, which may be somewhat misleading, a few special details of their anatomy and behavior recall the marmosets, howlers, and/or spider monkeys. Their geographical distribution is most like that of the marmosets. They occur throughout much of the Amazon basin in the widest sense and also in southeastern Brazil and adjacent areas, but they do not reach Central America or even the coasts of Colombia.

The subgroup has been revised by Hershkovitz (1963). He recognizes three species: *moloch, torquatus,* and (tentatively) *personatus*, each one of which can be divided into several subspecies.

Figure 20. Hostile facial expressions of titi monkeys. The two top drawings illustrate different subspecies of Callicebus moloch. *The two bottom show* C. torquatus.

C. moloch is the best known. Wild individuals of the species have been observed near Iquitos in Peru (W. G. Kinzey, personal communication; Mittermeier, personal communication), in the Meta region of Colombia at Barbascal near the town of San Martín (Mason, 1966 and 1968, Moynihan, 1966) and in La Macarena (Klein and Klein, op. cit.), in the Caquetá (personal observation), and in the Putumayo (by myself near Santa Rosa, by Le Nestour in other areas). Hernández-Camacho and Cooper describe some aspects of the biology of the species in Colombia in general. Several individuals from Colombia or adjacent parts of Amazonian Peru also were kept in captivity on Barro Colorado Island for a couple of years. All Colombian and northern Peruvian *moloch* are quite similar in coloration, largely gray or grayish brown above and rufous below (the scientific name refers to their "brazen" bellies). They belong to closely related subspecies such as *ornatus, discolor*, and *cupreus*. (The Caquetá animals are anomalous and may deserve special mention. Hershkovitz ignored them, simply because he thought that the species was absent from the area. This is surprising, for *C. moloch* is conspicuous around Valparaíso, one of the important towns of the *intendencia*. The individuals that I managed to see clearly, close up, in the Caquetá lacked the white stripe above the eyes that is typical of both *ornatus* to the north and *discolor* to the south. They could have been intermediates between one or both of the latter forms and *cupreus*, which occurs downstream, or representatives of an unnamed subspecies.)

All the *moloch* individuals that I saw anywhere in the field were inhabitants of dense vegetation, crowded and relatively low forests, thickets, and tangles. In this respect, the species might be supposed to resemble many of the other smaller ceboids. It differs from the others, however, in being usually or often linked to peculiar edaphic circumstances.

Individuals of the species frequently seem to prefer sites that are wet on the ground, poorly drained or even waterlogged.

At Barbascal in the Meta, they are found in narrow strips of gallery forest along streams and in isolated patches of woods, possibly relict and modified gallery, in the midst of savannahs (see also Thorington, 1967). These woods look like second growth, with trees scarcely reaching 20 meters above ground, but they may well be mature. The account of Hernández-Camacho and Cooper suggests that an association with gallery forest is common among the *moloch* of Colombia. The individuals near Santa Rosa in the Putumayo usually occur in depressions along creeks or in small valleys within extensive but rather heterogeneous forests far from existing human settlements. The trees of these valleys may average higher than the woods of Barbascal, but they tend to be shorter than those of the forests of adjacent slopes and ridges, and they are characterized by a profusion of vines and lianas, often very large ones. (The appearance of many of the areas inhabited by *moloch* in the Meta and near Santa Rosa is distinctive, almost diagnostic, somehow recognizable to the human eye in ways that are not easy to specify. The general effect is "messy." Part of the apparent disorder seems to be due to the scattering of twisted and often woody lianas leaning in all directions. Some of these reach heights of 10 to 15 meters with little visible substantial support. They may have grown up along or over old trees that have since fallen down and decayed without taking all their dependents with them.) The individuals seen by Le Nestour were farther up the Guamés drainage but confined to river bottoms. The Kleins found *moloch* in inundated terrain in the Macarena. One of the habitats in which I saw the species in the Caquetá was mostly low second growth forest, except for a few large trees, stretching along a narrow strip of very flat land between a broad river on one side and patches of bamboo and abandoned crop fields on the other.

I found another pair or family group of *moloch* in an even more extreme habitat near Valparaíso (really very near indeed, just outside the town limits). This was a medium-sized expanse of incredibly dense, almost solid, and low forest of small, thin, broadleaved trees and large bushes, hardly 7 meters high at its maximum. During the rainy season, the whole ground underneath was a morass of pools, streams, and deep pits of liquid mud. In the dry season, some of the pools and streams disappeared, but the area remained damp and the footing very treacherous. Naturally, the titis of this thicket averaged much lower than those of less stunted vegetation.

There are indications that other populations of the species may be equally "water loving." The type locality of one of the southern subspecies, *donacophilus*, is "*Dans les bois et parmi les roseaux qui bordent les rivières de la province de Moxos dans la République de Bolivia*" (d'Orbigny, 1836, quoted by Hershkovitz, 1963).

To my knowledge, no other ceboid has been seen in similar environments so frequently. *C. moloch* is the nearest thing to a real swamp monkey in the New World.

Individuals occasionally stray into habitats that are either atypical or unnatural. Even this can be revealing. The Kleins saw a few *moloch* in forests of foothills near the headwaters of streams flowing into the Rio Guayabero, one of the affluents of the Rio Meta itself. They describe these forests as low and "simplified." I should imagine that the simplicity refers to plant species composition or diversity, not to physical structures. A farmer in the Caquetá told me that the local *moloch* enter groves of imported Asiatic giant bamboo from time to time. Comparing his account with those of my informants in the Putumayo, it would appear that titis of this species occur in bamboo more frequently than the *Callimico* or *Saguinus fuscicollis* of the island in the Rio Guineo. The habit is of interest because bamboo is so forbidding to most animals. It is as difficult to penetrate as any plant formation in the world. Perhaps the titis can enter and exploit the formation because they are already adapted to the density of the native swamp and riverine forests and thickets.

Whatever the nature of their immediate surroundings, *moloch* individuals usually choose the thickest parts of the available vegetation for both feeding and resting. They may be even more pertinacious than tamarins and marmosets in this respect. In their own somewhat peculiar way, they are also among the most thoroughly arboreal or at least "nonterrestrial" of ceboids. I saw them on the ground only when they fell or were pushed during violent disputes.

The diet of the species is mixed, including berries, other fruits, insects (Coleoptera and Orthoptera), spiders, millipedes, and perhaps buds, but few or no large leaves (personal observation, and Hernández-Camacho and Cooper, op. cit.) The mix may be rather more vegetarian than in the diets of some other small monkeys, the tamarins and some or all of the marmosets.

This might help to explain peculiar aspects of spacing and mobility. *C. moloch* individuals can move very rapidly and with great agility, always quadrupedally, whenever they want to. But they do not seem to want to do so very often. They have relatively small territories, and do not range outside of them in the normal run of events. This must mean that their food is abundant and closely packed in appropriate habitats. Fruits are more likely to conform to this pattern than are arthropods. The owners of a territory usually can survey their boundaries from a single central point or small cluster of points. Much of their intraspecific territorial defense is vocal in any case (Moynihan, 1966). Thus, they have only occasional need to bestir themselves.

The contrast with some howlers may be emphasized by repetition. Both *Alouatta villosa* and *Callicebus moloch* are less active than all or most of their relatives, possibly because they both rely upon plant foods that are easily accessible. But their inactivity takes different forms. *A. villosa* groups have more or less large territories and usually move through them slowly. *C. moloch* have small territories, through which they move rapidly but infrequently. (Why there should be this territorial difference is another question, which cannot be answered yet.

Doubtless both sizes of animals and sizes of foods are involved.)

The ecological relations among *moloch* and various tamarins probably are complex. The distributions of these animals have puzzling gaps and limits. Some species may have been prevented from spreading widely and rapidly by physical barriers such as major rivers or by special preferences of one sort or another. Other factors of a more historical nature may have to be invoked to explain other constraints.

Consider the possible effects of *Callicebus moloch* upon *Saguinus fuscicollis* and vice versa. The two species are close neighbors in parts of the Caquetá (see also Chapter 4) and probably many other places. But the tamarins seem to be absent from the Barbascal area of the Meta, not very far to the northeast, although much of the local vegetation looks suitable for them. *C. moloch* individuals are abundant by contrast. Possibly they have excluded tamarins from the area. The proximate cause may be competition for space. The two species take rather different selections of foods. The diet of *moloch* probably is less like that of *fuscicollis* than is that of the Squirrel Monkey, with which *fuscicollis* coexists throughout all or most of its range. Yet the titis of Barbascal often move and rest in the particular tangles and trees that would be expected to be most attractive to tamarins.

Some of the forests inhabited by *moloch* in the Meta are farther from permanent bodies of water or slightly less dense than are those in which the species has been found in the Caquetá. The *moloch* niche appears to be relatively broad near Barbascal. This could be explained by supposing that individuals of the species arrived at a favorable time. They probably were not the earliest arrivals in the area (see comments in Moynihan, 1966), but they may well have come in before any tamarins. If so, the story may be decipherable. The first titis may have seized the opportunity to expand from their originally narrow base in dense riverbed or swamp vegetation into somewhat more open and slightly better drained habitats. When and if tamarins arrived on the scene at a later date, they may have been confronted by a blank wall. Titis are larger than tamarins. They may be more intimidating. Possibly tamarins could not insert themselves, physically, onto sites where titis were already sitting.

The assemblage of small primates on the island in the Rio Guineo may be the result of similar selection working in the opposite direction. Here it is *Callicebus moloch* that is absent from vegetation, part of the intermittently flooded forest, which looks eminently suitable; this in spite of the fact that the area is, or has been, easy to reach. (The river is narrow when not in flood; its course must be changeable; and there are populations of *moloch* on both sides.) The potential place or role of *moloch* on the island may have been taken by *Saguinus fuscicollis* or *Callimico goeldii* or, more probably, both.

S. fuscicollis and *S. graellsi* may have replaced or excluded one another from some areas. The latter species may also compete with either *Callicebus moloch* and/or *Callimico goeldii*. (One poorly drained forest near Santa Rosa was full of *graellsi* but lacked both *Callicebus* and *Callimico* as well as *fuscicollis*.) Competition for food should be more significant among tamarins than between tamarins and titis.

It would seem that interactions between the same species may have different outcomes in different areas, even in very similar environments.

Both Mason and the Kleins have suggested that *Callicebus moloch* and *Saimiri* also tend to be mutually exclusive. I do not think that the exclusion can be actively social. I would expect Squirrel Monkeys to be less attracted to *moloch* habitats than are tamarins (see also below). If there is any serious competition in this case, it could be for either food or space.

C. moloch individuals pay comparatively little attention to most potential predators. They act as if they were largely immune to attack. Perhaps their thickets and tangles usually are close and strong enough to be safe refuges. Certainly they sleep in their tangles rather than in holes.

Callicebus torquatus resembles *moloch* in general appearance and bodily proportions. It is only slightly larger and more deeply colored. Most of the populations of the species are rufous above and black below, with a whitish semicircular stripe on the throat or upper chest, and whitish hands in most forms.

The species is confined to part of the Amazon basin. Its geographical range, in the broad or loose sense, overlaps that of *moloch*. Still the term "overlap" may convey a misleading impression, for the two species have some different habits.

Hernández-Camacho and Cooper say that *torquatus* occurs in a variety of forests from gallery to rain forest in Colombia. A few individuals of the species that I saw near El Pepino, Rumiyaco, and Santa Rosa were in tall mature forest, or mixed forest with many tall and large trees, on well drained ground. The Santa Rosa animals were on the ridges and slopes rather than in the adjacent depressions. In all three areas, the *torquatus* seemed to prefer the same type of vegetation as many *Pithecia monacha*, and they tended to remain almost as high above ground. Other individuals studied by Kinzey (personal communication) in the Nanay River region of Peru were found in well-developed forest, usually on the tops of hills. They often moved and fed around 25 meters above ground, in the highest continuous canopy of the area but well below the tallest emergent trees.

Like so many other monkeys, individuals of *torquatus* are opportunistic in their choices of foods. They were seen to eat the pulp of palm nuts of the species *Jessenia polycarpa* in the Putumayo. These nuts are rather large. Hernández-Camacho and Cooper suggest that *torquatus* individuals take significantly fewer arthropods than do *moloch* individuals in the Guayabero region. Humboldt (1812) describes how a captive *torquatus* killed and ate birds. Kinzey gives more information on the natural feeding behavior of a few individuals along the Nanay. They took some insects, buprestid beetles and wasp larvae and perhaps galls. They also fed on some 59 recognized species of plants, of which 14 seemed to be important to them. They ate quite a lot of leaves and a miscellany of fruits and legumes (including *Inga*). Again the most favored food was the pulp of *Jessenia* nuts.

C. torquatus and *moloch* may not be segregated by their different habitat preferences at all places and times. Some of the available evidence is ambiguous. (Thus, for instance, Mittermeier found *moloch* to be quite abundant on "*terra firme*," supposedly never flooded, upriver from Iquitos; but this cannot be very far from the area in which Kinzey worked, and his impression is that the local *moloch* occur on lower ground than do the local *torquatus*.) Where and when the two kinds of titis should come into contact with one another, competition between them may be controlled or reduced by differences in diet. There are indications that *torquatus* individuals take larger items of food than do *moloch*, as well as a different assortment of species.

Perhaps partly as a consequence, *torquatus* seem to have substantial territories or home ranges, certainly larger than those of many *moloch*.

The *torquatus* of the Putumayo are also comparatively shy of man. They are particularly likely to freeze, become silent and motionless, when disturbed. They may do so more frequently, and for longer periods on the average, than any of the other ceboids that have been studied except *Cebuella*. They must be vulnerable because they are not very large and their immediate environments do not usually contain many tangles or thickets.

Obviously the *Callicebus* type is labile. It can adjust to different life styles with a minimum of physical change. This should be true of all generalized animals, but none of the other genera of New World primates provides a more striking example.

Spider Monkeys and the Woolly Monkey
(*Ateles*—including *"Brachyteles"*—and *Lagothrix*)

Woolly Monkeys of the genus *Lagothrix* are the least specialized members of this subgroup. They look rather like howlers at first glance. They are large and have prehensile tails, and
some of their body and limb proportions are similar to those
of *Alouatta* (see Figures 16 and 21), but the special resemblances between the two genera seem to be largely convergent.
Some of the basic traits of Woolly Monkeys are quite distinct
from those of howlers. Woollies have large and complex brains
and lack the peculiar vocal apparatus of the howlers.

They occur in the valley of the upper Magdalena River in
Colombia, throughout the upper Amazon basin, and on the
foothills and eastern slopes of the Andes, reaching as high as
3000 meters above sea level, much higher than most other
ceboids. There are several forms within this range. They have
been revised by Fooden (1963), who recognizes two species:
lagothricha, with four rather well marked subspecies, and a
monotypic *flavicauda*. His ranking of the latter may be ques-
tioned. His account does not suggest that *flavicauda* is much
different from the other forms than they are from one another,
and its range does not overlap any of theirs. Thus, the most
logical solution would seem to be to include all the forms in a
single species, for which *lagothricha* is the oldest and, therefore, the best name.

They all have simple color patterns, mostly black, or gray,
or brown, or mahogany red.

Wild individuals of the species have been observed by biologists in many areas. Unfortunately, most observations have
been brief and "spotty," made by students of different backgrounds and interests, and often as ancillary by-products of
studies of other subjects. The natural history of Woolly Monkeys is not as well known as it deserves to be, or as one might
have expected of such large and conspicuous animals. Probably
the best ecological data are in Hernández-Camacho and Coop-

er's general survey of Colombian primates as a whole. Durham (personal communication) observed some population in Peru. I myself saw several groups of the species in the Caquetá. These animals may have been representatives of the subspecies *lugens* or intermediate between the latter and the nominate subspecies.

All Woollies seem to be inhabitants of humid forests. In the lowlands of the Caquetá and many other lowland areas, they occur in tall, rich and diverse, mature, or at least old, forests. The individuals that I observed were found in well-drained forest, often on slopes. Hernández-Camacho and Cooper say that other individuals occur in seasonally flooded forests, gallery, palm forest (the *Mauritia flexuosa* association), and "cloud forest" in the mountains. They usually avoid young or obvious second growth (although not, according to Durham, banana plantations).

They wander through a variety of heights above ground. Durham says that they are characteristic of midcanopy levels. Hernández-Camacho and Cooper say that they occur most frequently just beneath the canopy, occasionally going up into emergent trees. In the lowland Caquetá, they move through both medium and large trees 10 to 35 meters above ground.

Many of their locomotory patterns are like those of howlers in form, but they show a more definite or greater tendency to brachiate, and they move more frequently and rapidly on the average, than the *Alouatta villosa* of central Panama. They are quite capable of sweeping through the trees with considerable speed, if not with the extreme velocity, grace, and acrobatics of sakis or spider monkeys.

They are primarily, perhaps almost exclusively, frugivorous. The individuals that I observed in the Caquetá took many different kinds of fruits, usually of medium to large size. Cruz Lima (op. cit.) says that the Woollies of the Brazilian Amazon are frugivorous but does not give details. F. C. Lehmann (personal communication) observed individuals feeding on figs and

Figure 21. *Adult male Woolly Monkeys: top, an apparently purely hostile posture; bottom, an ambivalent posture, partly hostile and partly friendly.*

the fruits of *Cecropia* in the district of Moscopán in Colombia, on the eastern slopes of the central cordillera, east of the Volcan de Puracé, around 2000 meters above sea level. Fooden, noting that the teeth of adults usually are worn down, suggests that many of the fruits taken are hard-shelled. There is some confirmation in field reports. Durham says that the preferred foods of Peruvian Woollies are fruits and buds of *Ficus* spp., "Tinta"(?), *Cecropia*, and *Palmetto* nuts and shoots. Hernández-Camacho and Cooper say that populations of the eastern plains of Colombia take quantities of palm fruits, particularly *Mauritia flexuosa, Attalea regis* (syn.: *Maximiliana elegans*), and *Jessenia polycarpa*.

Both Woolly and spider monkeys have suffered greatly at the hands of man. Wherever they occur, they are thought to be delicious. They are hunted even in areas where no other monkeys are supposed to be suitable for the pot. Rather surprisingly in these circumstances, neither the Woollies of the Caquetá nor those of southeastern Peru have yet become very shy of man. Perhaps the species had almost as much to fear from birds of prey as from predatory mammals under natural conditions. Lehmann (personal communication) saw a Woolly Monkey, probably an adult female, being carried (with difficulty) by a Harpy Eagle, *Harpia harpyja*, in the Vaupés. The Woollies of Moscopán also react vigorously to a slightly smaller eagle, *Oroaetes isidori*. They utter alarm calls and drop down to low branches whenever one of these birds flies overhead (Lehmann, 1969). According to Hernández-Camacho and Cooper, a third eagle, *Morphnus guianensis*, similar to *Harpia* in appearance but more like *Oroaetes* in size, has the popular reputation of being another specialist upon Woollies. (One would not, offhand, have expected birds of prey to pay much attention to such a large monkey. The fact that they do bother to do so may reflect either the unwariness of the species or its exceeding attractiveness.)

There are hints of other causes of mortality. Woollies of the eastern plains of Colombia collected by Hernández-Camacho were heavily infested with filarid nematodes.

The only demographic figures come from a border region. Durham suggests that the population densities of *Lagothrix* in southeastern Peru range from 14 to 35 per square kilometer.

Spider monkeys are lighter, lankier, more elegant and agile than Woollies. Their locomotory movements have recently been described in great detail by Mittermeier and Fleagle (in press). They brachiate frequently and superbly. Presumably in correlation, their arms are elongated and their hands have become modified, transformed into suspensory hooks. Their thumbs have disappeared or been reduced to vestiges, and the remaining fingers are long and permanently curved. When they cannot or do not use their arms, as on the ground and some large branches, they walk quite well on their legs alone. All this is reminiscent of the small brachiating apes of southern Asia and the Greater Sunda Islands, the gibbons and Siamang of the genus *Hylobates*. But the latter have sacrificed or de-emphasized other kinds of movements, which is not the case with their New World analogues. Spider monkeys can still run and leap quadrupedally with the body more or less horizontal in many circumstances. They use the prehensile tail as extra support whenever possible. Thus, they would appear to have remained more adaptable than the gibbons, an impression that is reinforced by their extreme flexibility of social behavior, especially in the organization of social groups (see below and Chapter 4).

There are two main kinds of spider monkeys. One is the so-called Woolly Spider Monkey, *arachnoides*, of the forests of southeastern Brazil. It is often placed in a separate genus by itself, "*Brachyteles*," and has sometimes been thought to be intermediate between *Lagothrix* and the more typical spider monkeys of the genus *Ateles*. This may be going too far. The basic characters of *arachnoides* seem to be essentially completely spider monkeylike. It certainly is a good species. Possibly it should be recognized as a subgenus (this is the policy followed by Fiedler, 1956), but nothing more.

Woolly Spider Monkeys seem to have become very rare in recent years. To all intents and purposes, we are ignorant of their habits in the wild.

The other spider monkeys pose several taxonomic and systematic problems. They range throughout most of the lowlands of tropical Central and South America, except southeastern Brazil, and into the foothills of the Andes and the mountains of Central America and Mexico. There are many distinct forms. They differ appreciably in size, and also in color and pattern. The variation involves both the bare skin of the face and the pelage. Some forms have black faces. In others, the faces are partly or wholly light or even bright red. The pelage may be black, sometimes marked with white or golden yellow, or gray, or largely rufous or golden all over. Several forms are illustrated in Figures 14, 22, 23, and 24. The whole assemblage has been revised by Kellogg and Goldman (1944), who divided it into four species—*paniscus, belzebuth, fusciceps*, and *geoffroyi*—but suggested that the number might be reduced in the future. This seems eminently reasonable. All the forms are primarily or exclusively allopatric. Two of the forms that are most different in appearance, a large black "*fusciceps*" type and a small red "*geoffroyi*," are known to have hybridized quite frequently in Panama in an area where their ranges approached or met (again see Chapter 4). If this could happen, there is no reason to suppose that any of the other forms have developed effective genetic, physiological, or behavioral isolating mechanisms. Unless and until some additional and contradictory evidence is obtained, they may all be placed in the same species. This must be given the scientific name of *paniscus* and might be called "the Common Spider Monkey" in the vernacular. For practical purposes, however, it will be convenient to continue to use the "specific" names of Kellogg and Goldman as appropriate designations for particular groups of subspecies.

Figure 22. Common Spider Monkeys: top and center, Ateles "geoffroyi"; *bottom,* A. "belzebuth."

Figure 23. Red Common Spider Monkey, "geoffroyi," maneuvering on palm fronds. It is easy to see why such fronds would be difficult for some other monkeys. Photo by O. Linares.

Figure 24. *Resting or alarm posture of Red Common Spider Monkey in* Psidium *tree. Even an animal as large as this can be cryptic in the right circumstances. Photo by O. Linares.*

Common Spider Monkeys have been observed in the field under both natural and seminatural conditions. Carpenter (1935) studied a population of "*geoffroyi*" in the Cotó region of western Panama (Chiriquí). P. Wagner (1956) describes other "*geoffroyi*" in Mexico. Eisenberg and Kuehn (1966), the Hladiks and their collaborators (works cited above), Richard (op. cit.), and others, have studied different aspects of the biology of a few semitame "*geoffroyi*" individuals, mostly hand-reared, reintroduced into the forest of Barro Colorado Island. Husson (op. cit.) has brief notes on individuals (*paniscus s.s.?*) in Surinam. I myself saw a few "*belzebuth*" in the Caquetá. Klein (1972) and Klein and Klein (op. cit.) have more data on the "*belzebuth*" of the Macarena. Durham saw groups (cf. "*paniscus*" *s.s.?*) in southeastern Peru.

It seems evident that Common Spider Monkeys are primarily or perhaps originally inhabitants of old, tall, and spacious forests. They also seem to be overwhelmingly frugivorous in all or most areas. The "*belzebuth*" of the Caquetá move and feed in many of the same sorts of places as the local Woolly Monkeys. So do the closely related and similar looking individuals of the Rio Guamal in the Meta (Lehmann, personal communication). Durham says that Peruvian spider monkeys live and feed within "pockets" of *Inga, Lupuna,* and *Palmetto,* or *Ficus* and *Lupuna,* or *Palmetto* and *Ficus.* The Kleins state that the animals of the Macarena occur in special patches of vegetation, some or all of which may be less rich or diverse than most old or mature lowland forest (see also below), and prefer fruits of such trees as *Hyeronomia, Brosimum* sp., and *Calophyllum* sp. Hernández-Camacho and Cooper say that the spider monkeys of Colombia as a whole occur in remnant and degraded forest as well as full rain or monsoon forest, and feed on *Spondias* and *Jessenia* in addition to the fruits cited above. They further note that it is not uncommon for these animals to forage on the ground in some circumstances.

There are, as usual, more detailed calculations in the papers of the Hladiks et al. The spider monkeys of Barro Colorado take approximately 20% leafy material and 80% fruits when they are moving around the forest. Eight per cent of the material ingested in the forest is provided by the following plants: *Mangifera indica, Spondias monbin, Annona spraguei, Quararibea asterolepis, Doliocarpus* sp., *Dipteryx panamensis, Inga* spp., *Trichilia cipo, Cecropia* spp., *Ficus* spp., *Poulsenia armata, Virola panamensis, Psidium guajava, Astrocaryum standleyanum, Socratea durissima, Tocoyena pittieri, Citrus* sp., and *Chrysophyllum cainito.* This breaks down to 7.4% protids, 4.9% lipids, 33.7% reducing glucids, 11.0% cellulose, and 43.0% complementary fractions. The proportion of protids is comparatively and perhaps abnormally low. Possibly it is supplemented by foods stolen from the cages and pens of laboratory animals (and even the laboratory kitchen). Be that as it may, the spider monkeys certainly show a preference for the most nutritious fruits, not only *Ficus* spp. (which they suck instead of eating entire like the howlers) but also such forms as *Psidium guajava* (not native to the island) and the palm *Astrocaryum standleyanum.*

Groups of both the Woolly Monkey and the Common Spider Monkey have relatively very large territories or home ranges, which they cover more or less rapidly by fits and starts.

The ecological relations between the two species are puzzling. Where they occur in the same areas, they seldom associate with one another (see Chapter 4). They probably take many of the same fruits with different frequencies. Possibly each takes some fruits that the other ignores. Their relative abundance probably differs from area to area. Common Spider Monkeys appear to be less numerous than Woolly Monkeys in southeastern Peru (Durham suggests that the local populations of spider monkeys in the latter region may vary from 6 to 24 per square kilometer), but this was not necessarily true of all the sites in the Caquetá that I visited. The two species also seem to be spread unevenly within their ranges. There are clusters of individuals and groups in and around concentrations of favored food sources, and few or no individuals or groups between concentrations. (Mature humid tropical forest is heterogeneous in composition. At first sight, it may look equally and similarly heterogeneous throughout. This appearance is deceptive. Even apart from obvious accidents such as windfalls, species of plants of tropical forests are distributed in a noncontinuous manner. Concentrations of particular plants must affect the distributions of all New World primates. The general effect has been remarked upon more frequently in the case of *Lagothrix* and *Ateles* than in the cases of most other high forest forms.) Incomplete as they are, the available accounts would suggest that populations of the Common Spider Monkey are more unevenly distributed, more clustered or patchier, than are those of the Woolly Monkey where the two species are sympatric. The Woolly Monkey may be (or may have been) more dominant than the Common Spider Monkey in optimal habitats in such areas. But the *Ateles*, of course, also occur in other regions. This could be because spider monkeys can tolerate more degraded or less rich forests than can Woolly Monkeys, and/or because they come down to the ground more frequently and can thus pass from forest to forest more easily. In areas where Woollies are absent, Common Spider Monkeys may be more abundant or more evenly spread than in areas where Woollies are present.

Presumably both Woolly and Common Spider Monkeys take some larger and/or harder fruits than do such forms as *Pithecia monacha* and *Callicebus torquatus*.

Lacking thumbs, spider monkeys are not good manipulators. They cannot, therefore, explore or control some features of their surroundings as intensively or efficiently as can *Cebus* spp., which have hands that can do more different kinds of things. (It might be mentioned, however, that the *Ateles* of Barro Colorado and the Caquetá do not seem to suffer from skin parasites to anything like the same extent as Panamanian *Alouatta villosa*.) Common Spider Monkeys may be the most adaptable of ceboids as far as their locomotory patterns and some of their social reactions are concerned. Possibly they alter the frequency and nature of their social contacts quite rationally, even consciously, depending upon conditions. They have the same large and complex brains as Woolly Monkeys. They appear to be very intelligent (see also Polidora, 1964). Nevertheless, they are not flexible in all respects. Their reactions to many objects are comparatively stereotyped. Like all the New World primates considered so far, they have never been seen to use tools either in captivity or in the wild.

They definitely do not sleep in holes in trees. Some early accounts (cited in Cruz Lima, op. cit., and Cabrera and Yepes, op. cit.) would suggest that they were rather contemptuous of, occasionally aggressive toward, potential predators at one time. They may have been the "small hairy Indians with tails" encountered by the first Portuguese explorers in the Amazon.) Now they are being hunted with modern firearms in many areas and have become moderately to very shy of man. In the Caquetá, the local spider monkeys seem to be warier than Woolly Monkeys.

The Squirrel Monkey (*Saimiri*)

This genus appears to be almost perfectly intermediate between *Saguinus* and *Cebus* in some features of behavior and gross morphology. The resemblances may or may not be valid indications of close phylogenetic relationships. They certainly reflect broad ecological contact or overlap.

Squirrel Monkeys occur in the guianas and throughout most of the Amazon basin down to Bolivia, and also in western Panama and southern Costa Rica. They seem to be absent from eastern Panama and northern Colombia. The taxonomic history of the genus is reviewed in Cooper (1968). There may be only one species, *sciureus*. The isolated Central American population seems to agree with some or all of the South American populations in most aspects of behavior and ecology. All the Central American animals may be included in a single subspecies, *oerstedii*. It is obvious that there are several subspecies (and possibly morphs) in South America, but no one seems to know exactly how many, or what subspecific names should be applied to the animals of many particular areas.

The various forms do not differ much in appearance. They are more or less grayish olive on most of the upper parts, largely whitish below, with a wash of yellow or orange on the limbs, black-tipped tails, and a conspicuous pattern of black (on the muzzle) and white (around the eyes) on the face (see Figure 25). They are quite small, only slightly larger than the largest tamarins. They must be classed as quadrupedal springers. Their limbs are subequal in length. Their thumbs are well developed and partially opposable. They can manipulate objects with moderate dexterity. (Chippendale, quoted in Kortlandt and Kooij, 1963, reports that he saw a Squirrel Monkey use a stick as a tool to move fruit along the ground.) Their tails are slightly prehensile, but are used for support only in an accessory role, and then only infrequently. Their brains are larger than those of most other primates of comparably small body size, but apparently less highly developed than those of Woollies or spider monkeys.

I have seen Central American Squirrel Monkeys near Puerto Armuelles and on the Burica Peninsula, along the Pacific coast and the frontier between Panama and Costa Rica, and South American animals in the Meta, near San Martín, in the Caquetá, and in the Putumayo near Santa Rosa. Hladik and Hladik (op. cit.) studied the same Central American populations, and Thorington (1967 and 1968b) the same population in the Meta. Hernández-Camacho and Cooper have notes on the species throughout Colombia. The Kleins observed several groups in La Macarena; Durham considers the status of the species in southeastern Peru. DuMond (1968) and Baldwin (1968, 1969, and 1971) have published several papers on Squirrel Monkeys in semifree-ranging conditions in Florida, and Baldwin and Baldwin (1971 and 1972b) have described aspects of natural populations in Panama, Colombia, Brazil, and Peru.

The species as a whole occurs in a very great diversity of habitats. Hernández-Camacho and Cooper mention gallery

Figure 25. Squirrel Monkeys.

forest, low sclerophyllous (dry) and hillside forests, palm forest
(the *Mauritia flexuosa* association again), and both seasonally
flooded and nonflooded rain forests. Durham seems to imply
that the southeastern Peruvian populations are concentrated in
or confined to gallery forests (possibly they are more widespread but difficult to see in other environments of the region).
In both Chiriquí and the Caquetá, Squirrel Monkeys are most
characteristic of humid lowland areas where the climax vegetation is, or would be, tall and spacious forest. In these areas,
they take advantage of almost everything, not only whatever
remains of the climax, but also most of the new artificial or
man-made habitats, all kinds of edge and second growth,
hedges on the outskirts of (or even within) towns, crop fields,
plantations of fruits (citrus, bananas, plantain), and stands of
imported trees such as teak. Remarkably, they are often abundant in all these types of vegetation, from the oldest to the
youngest. They seem to flourish in more kinds of habitats than
any other New World primate with which I am familiar.

Wherever they occur, they range from the tops of the tallest
available trees, 50 or more meters above ground in some cases,
down to the ground itself. Their trips to the ground may be
occasional but would appear to be voluntary (viz. the contrast
with such forms as *Callicebus moloch* and *Alouatta villosa*).
They move along branches of all sizes and in spiny palms as
well as easier trees. They are very active and restless, passing
through many widely different types and levels of vegetation
in rapid succession without hesitation.

They seem to be restricted in only three ways. They refuse
to remain in really open country, grassland, for any appreciable length of time, although they will cross it to get from one
patch of forest or scrub to another. They do not usually enter,
may even tend to avoid, the waterlogged swamp forest and
thickets that are the homes of many *Callicebus moloch*. They
seldom occur at high elevations above sea level. There is an
apparently reliable sight record of a few individuals around
900 meters on Cerro Campana in Panama (F. A. McKittrick

and N. G. Smith, personal communication), but this was extremely exceptional. All the animals observed by Durham were below 300 meters. The species certainly does not go as high in mountains as *Lagothrix* or *Ateles*. It may also be limited to lower altitudes than *Saguinus fuscicollis, Pithecia monacha, Callicebus torquatus,* and *Cebuella*. It is absent from the environs of El Pepino and Rumiyaco now. Local settlers say that it never did occur there regularly, although a few strays used to appear at rare intervals when there was more natural vegetation left.

Most Squirrel Monkeys take a great variety of foods. Like tamarins, they eat many small fruits and still look for insects and other small arthropods. They pilfer the plantations that they visit. Among the food items listed by Hernández-Camacho and Cooper are fruits of *Cecropia* spp., *Ficus* spp., *Euterpe* spp., Rubiaceae, *Campomanesia* sp., plus berries, insects, and spiders. Mittermeier (personal communication) and the Kleins add frogs (*Hyla* and *Sphaenorhynchus* spp.) and frog eggs. A few data collected by the Hladiks in the vicinity of Puerto Armuelles indicate that the *Saimiri* of this area tend to take more fruits than do the *Saguinus geoffroyi* of central Panama. Perhaps 20% of their diet consists of animal foods, leaving 80% for vegetable material. These percentages are reminiscent of *Cebus capucinus* on Barro Colorado. *Saimiri* and *Cebus* may be as nearly omnivorous as any American primates.

The Kleins estimate the density of Squirrel Monkeys in La Macarena as 50 to 80 individuals per square mile. It was my impression that they may be even more abundant in parts of the Caquetá and the Burica Peninsula. Sanderson (op. cit.) and others have claimed that Squirrel Monkeys are the most numerous of all ceboids. If so, it is because they have the broadest range or combination of food and habitat preferences.

They do not seem to sleep in holes, in spite of their small body size. Probably there are not enough holes to accommodate them all. As would be expected, they are wary of potential predators. Their actual defense mechanisms are both simple and somewhat peculiar, but obviously effective enough; they are discussed in Chapter 4.

Capuchin Monkeys (*Cebus*)

These are the most typically monkeylike of all ceboids. They are the animals that used to accompany organ-grinders in the streets of New York and Boston. They certainly are a climax of ceboid evolution.

They are quite diverse, and it is definitely known that they include several species. The most convincing revision of the subgroup is by Hershkovitz (1949). He identifies four species, to which he assigns the names *capucinus, albifrons, nigrivittatus*, and *apella*. Some of this nomenclature may be controversial, although probably not more so than some of the substitutes that have been proposed, but the reality and distinctiveness of the four taxa are not open to doubt.

All forms are moderately large. The first three species are comparatively lightly built, slender, and long limbed (least so in *nigrivittatus*). *C. capucinus* occurs in the southern part of Central America and along the Pacific coast of Colombia and probably Ecuador, in the lowlands and up to 2100 meters in the western cordillera of the Andes. It is distinctively colored, largely black with a whitish or pale buffy face and chest (the reason for the Latin name is self evident). Both *albifrons* and *nigrivittatus* show considerable geographic variation in color, but always tend to be various shades of brown. The former species has a peculiar distribution in the valley of the upper Amazon, along the banks of the upper Orinoco, the Lake Maracaibo drainage and some areas of northern Colombia, the island of Trinidad, and (again) the Pacific coast of Ecuador. *C. nigrivittatus* occurs in the Guianas and adjacent Venezuela and Brazil. *C. apella* is a stockier animal. It has a very wide distribution on the eastern slopes of the Andes, up to 2700 meters, and throughout much of the Amazon basin to southern Brazil, Paraguay, and northern Argentina. Many subspecies have been described. They also are brown, but they differ among themselves in intensity and tone of coloration, and in the development of tufts of elongated hairs on the forehead and temples. *C. apella* is illustrated in Figure 26, *albifrons* in Figure 27, and *capucinus* in both Figures.

Figure 26. Capuchins: from top to bottom, a hostile facial expression of an adult Cebus capucinus; profile of an adult male C. apella, showing the form of the forehead tufts in one subspecies; a locomotory intention posture of the same male apella; and a highly aberrant "quirk" of a juvenile capucinus.

All capuchins are quadrupedal springers. They have opposable thumbs and greater manual dexterity than other kinds of New World primates, plus prehensile tails, which are used to greater and more varied effects than are those of other forms —to grasp, carry, or even play with separate objects, as well as to support the body and aid in locomotion.

Wild *capucinus* have been studied at length. Oppenheimer (1968) analyzes many aspects of the life of a few groups of the species on Barro Colorado Island. The Hladiks and their collaborators discuss the foods and feeding of some of the same individuals and others at nearby sites on the mainland. Oppenheimer, the Hladiks, I myself, and several other students have made briefer observations in somewhat more remote areas such as the Burica Peninsula, the neighborhood of Puerto Armuelles, and the Darien in eastern Panama. Hernández-Camacho and Cooper have notes on the species in Colombia. Freese (op. cit.) reports on a far western population in the Santa Rosa park of Costa Rica.

In Panama, the species seems to prefer old forest in humid areas on well drained ground, but it also extends into mangrove and a variety of second growth habitats. In Colombia, it occurs in both primary and secondary forests, including highly degraded remnants of a few large trees and palms (particularly *Scheelea magdalenica*) in the midst of pastures. The more marginal Costa Rican animals occur through low deciduous forest, 5 to 7 meters high, in addition to tall evergreen forest and mangrove. Everywhere, individuals of the species range from the canopy downward. They resemble *Saimiri* in this respect, but they are too heavy to move along the most slender branches like the smaller species. They also come down to the ground, to the forest floor and for occasional excursions into or through open country, more frequently than Squirrel Monkeys or any other ceboids except *Cebus apella* and (probably) the other members of the same genus. Capuchins certainly must be considered to be primarily arboreal; they spend more time in trees than on the ground. Nevertheless, they have con-

Figure 27. Capuchins: top, two relaxed postures of Cebus albifrons; *bottom: a juvenile* C. capucinus.

verged toward some of the terrestrial Asiatic and African primates, such as the forest-inhabiting macaques and baboons, and they are the only American monkeys to have done so.

Most groups of *capucinus* have rather large territories and move through them rapidly. Individuals are extremely opportunistic in the course of this journeying. They will investigate anything that looks as if it might conceivably be edible. They will strip off bark and bite or dig into branches, shoots, and trunks, partly in search of insects (see below) and partly in the hope that the pith or some other tissue will prove to be palatable. They also bite into the majority of the fruits encountered. Many of these turn out to be unsuitable and are rejected immediately, but the number and diversity of those accepted is still impressive, larger than in the case of *Ateles* (to say nothing of *Alouatta*). Sometimes fruits are bitten and then collected later. Possibly the initial bites accelerate the process of ripening. Hladik et al. estimate that the diet of the *capucinus* that they studied was composed of 20% animal prey, 15% leafy and miscellaneous vegetable material, and 65% fruits. This breaks down to 14.4% protids, 15.8% lipids (very high), 26.3% reducing glucids, 7.6% cellulose, and 36.0% complementary fractions. The extremely catholic nature of the tastes of *capucinus* is revealed by the following list of plants, which provided (only) 60% of the vegetable material seen to be ingested: *Anacardium excelsum, Mangifera indica, Spondias monbin, Annona spraguei, Dendropanax arboreus, Ochroma limonensis, Cordia nitida, Doliocarpus* sp., *Rheedia edulis, Rheedia madruno, Lacistema aggregatum, Gustavia superba, Clitoria arborescens, Dipteryx panamensis, Inga* spp., *Miconia argentea, Mouriria pareiflora, Trichilia cipo, Cecropia* spp., *Ficus* spp., *Olmedia aspera, Musa sapientum, Virola panamensis, Psidium guajava, Desmoneus* sp., *Scheelea zonensis, Alibertia edulis, Faramea occidentalis, Pentagonia pubescens, Randia armata, Tocoyena pittieri, Zanthoxyllum* sp., *Cupania latifolia, Chrysophyllum cainito, Apeiba aspera, Hybantus anomalus.* This should be compared with the shorter lists of plants that make up higher proportions of the diets of *Alouatta villosa* and *Ateles "geoffroyi."*

There are other interesting aspects of the feeding behavior of *capucinus*. Unlike howlers or Common Spider Monkeys, *capucinus* individuals seldom or never gorge on a single kind of fruit for minutes or hours on end. Such favorites as they have are rather distinctive. The palms may be cited as an example. It has already been mentioned that the spider monkeys of Barro Colorado are devoted to *Astrocaryum standleyanum*. In the same general area, the capuchins tend to choose *Scheelea zonensis* instead (it must be the source of much of the fats in their diet). Capuchins take some larger items than *Saimiri* or *Saguinus*. They also have some peculiar techniques for obtaining animal prey. Tamarins and Squirrel Monkeys hunt largely by sight. They try to catch whatever insects and other small arthropods may be visible, exposed on the surfaces of the vegetation. Apart from this, they only poke among clusters of leaves and unroll curled leaves, looking for pupae. *C. capucinus* individuals do the same, but they also extract grubs from within the branches and trunks they bite into or tear apart. They sometimes look as if they could detect the presence of grubs by smell or sound. Doubtless their superior manual dexterity helps in breaking up branches. It may even have originated as an adaptation to subserve this function. (Capuchins do not usually dig uniform or regularly spaced holes when they bite into wood. Their assaults are more varied and forceful than the rather dainty drilling efforts of Pygmy Marmosets, and, obviously, they have a different purpose.)

Hernández-Camacho and Cooper add that the *capucinus* of coastal areas of both Colombia and Panama have been seen to feed on oysters, which they open with stones.

Capuchins are so active that they are difficult to count in the field; Freese suggests that there may be something like 250 to 350 *capucinus* individuals in the 100 square kilometers of the Santa Rosa park. This may be the order of magnitude of population densities of the species in many areas.

Cebus apella has been studied extensively rather than intensively. There is much information on Colombian populations in Hernández-Camacho and Cooper and in the Kleins' paper on La Macarena. I saw many individuals in the Caquetá and a few in the Meta. Thorington (1967) made observations at the same site in the Meta. Husson (op. cit.) and Fooden (1964) have brief notes on the species in Surinam. Kuhlhorn wrote several papers (e.g., 1939 and 1943) on the local form(s) of the Matto Grosso, south of the Amazon. Some characteristics of other southern populations are cited in Krieg (op. cit.) and Causey et al. (1948).

The species may be even more adaptable than *capucinus*. In Colombia, it seems to occur in all humid forests, primary, secondary, gallery, palm, and cloud. It may well occur in second growth relatively more frequently than other species of the genus. It is often quite abundant (the Kleins estimate the density of *apella* in the Macarena as 15 to 25 per square mile). Individuals move at all levels in trees and also come down to the ground. They eat both vegetable and animal materials. They use the same techniques as *capucinus* to obtain insects and grubs. Among the fruits taken, according to Hernández-Camacho and Cooper, are those of several palms (*Scheelea* sp., *Euterpe* sp., *Jessenia polycarpa, Syagrus inajai*) and other forms such as *Monstera* sp., *Spondias monbin, Insertia* sp., and *Pouruma* sp. Many populations of *apella* still occur in the near vicinity of human settlements. They are notorious for raiding crop fields and plantations. The common name of the species in Colombia and several other parts of South America is "*maicero*," maize-eater.

Many of the palm nuts taken are exceedingly hard, which leads to rather spectacular behavior in the field. Groups of *apella* in the Caquetá are almost constantly accompanied by the sound of loud knocking. They can be detected and identified by this noise at very considerable distances (long before they can be seen, or before any other noises, such as the crashing sound of their running and leaping, can be heard). Some of the knocking may be preliminary to breaking branches in the search for grubs, but most of it is the pounding of nuts against hard surfaces, obviously in the hope of smashing the shells to get at the nutmeat inside. Knocking sounds are much more characteristic of *apella* than of *capucinus*. *C. apella* individuals may take more nuts than any other ceboid, search for them more continuously, and/or concentrate upon those with the hardest shells. (In the Caquetá, at least, other monkeys such as *Pithecia monacha* and *Lagothrix* probably do not take anything too hard to be opened by biting.)

The smashing of nuts by pounding may be difficult—almost as close to tool using as the opening of oysters. *C. apella* must have as many occasions for complex manipulation as *capucinus*. It is evident, in fact, that both species manipulate frequently enough for the habit to have become "generalized" in some sense. Captive *capucinus* and *apella* tend to explore and play with all the objects they can lay their hands on even when they are not hungry. They are very prone to take things apart. They seem to have a passionate desire to discover what is inside or behind anything that can be pulled or plucked or dismembered. They can also put familiar and strange objects together, sometimes in quite surprising ways, when the spirit moves them.

The other two species of *Cebus* show similar tendencies in captivity.

The peculiar use of the tail by members of the genus may be an "extrapolation" of the drive to manipulate. Certainly capuchins can employ their tails as effective supplements to, or substitutes for, their hands in a wide variety of circumstances. Wild *capucinus* can carry food in their tails. I have seen a dozen adult and juvenile *apella* in the Lima Zoo go through even more complex maneuvers. They threw a pink handkerchief seized from an incautious visitor, back and forth among themselves for over an hour. The handkerchief was caught approximately equally frequently by the hands or in the tip of the tail. Every once in a while, it was thrown by the tip in a sort of sideways "undertail" slow lob.

This capacity for experimentation explains the brilliant success of capuchins in some types of psychological tests in the laboratory (see, for instance, Klüver, 1933). Individuals may react stupidly to some stimuli. Their attention span is apt to be short, and they are more easily bored than any other primates with which I am acquainted. (Their "fickleness" must be adaptive in nature. Probably because they often have to work for food that is not directly visible, it may not be worthwhile for a capuchin to spend too much time excavating a particular point on a branch or pounding a particular nut. If the object or site does not yield a reward within a few minutes, there may be no reward present or attainable. It may be most efficient, to conserve time and energy, to move on promptly to investigate another possibility.) With all their limitations, however, capuchins can cope with some classes of stimuli with remarkable ingenuity, almost unparalleled below the level of man himself.

There is not much to be said about *albifrons* and *nigrivittatus* in the wild. They would appear to be less dominant as well as less widespread than *apella*. According to Husson, the Surinam population of *nigrivittatus* inhabits hill forests, when it replaces *apella*, which prefers the coastal lowlands. A troop of *nigrivittatus* observed by Oppenheimer and Oppenheimer (1973) near the Rio Guarico in Venezuela fed at a large fig tree (*Ficus pertusa*) on several consecutive days. Hernández-Camacho and Cooper say that *albifrons* occurs in both primary and secondary forests in Colombia, and feeds on *Inga* sp., *Bellucia axinanthera*, and *Euterpe* sp. It certainly overlaps *apella* in the Caquetá. Here it occurs in old or mature forest on well drained ground. I saw the members of one group feeding on small cherrylike fruits. Overlaps between the same two species may be less extensive in the Putumayo. Local Indians near Rumiyaco say that *albifrons* inhabit old forest on the surrounding hills where *apella* is rare or absent now. I also saw many *albifrons* near Santa Rosa, where *apella* is definitely absent, in both waterlogged forest of no great height and in taller forest on well drained ground. In both habitats, they were behaving much like *apella* elsewhere. Some of their troops were accompanied by knocking sounds. Perhaps *apella* and *albifrons* compete so strongly that they are often incompatible. This might help to explain the scattered distribution of *albifrons* along the northern edges of the range of its rival.

All species of *Cebus* are wary of man, even, or especially, when raiding plantations. They are so very alert and circumspect that it is difficult for a human observer to determine their responses to other potential predators. They can hardly be vulnerable to anything except some carnivora, large hawks and eagles, and a few snakes. Hernández-Camacho and Cooper note that Colombian *apella* seem to be afraid of the eagles *Harpia harpyja* and *Spizaetus ornatus*. They also saw a group of *apella* chased by the large fisherlike Tayra (*Eira barbara*). Some of the *nigrivittatus* observed by the Oppenheimers flinched or dodged when birds of prey flew overhead. None of the capuchins is known to sleep in holes. *C. apella* and at least one population of *albifrons* would appear to have evolved a "mutual defense pact" with *Saimiri*. Again see below.

chapter four

SOCIAL RELATIONS AND ORGANIZATIONS

Social behavior in the broadest, and perhaps only logical, sense includes all interactions between or among individuals of the same or different species, whether friendly or unfriendly, gregarious, sexual, aggressive, parental, or other.

Intraspecific Behavior

Each species of primates has its own internal social arrangements. There have been several attempts to classify and rank such arrangements; see, for instance, Crook and Gartlan (1966) and Eisenberg et al. (1972). Some of the proposed typological refinements do not seem to me to be very pertinent to the ceboids. These animals have two extremes of social organization: the small "nuclear" family, and the large troop. (All statements in this and the next chapter are based upon personal observation or a consensus of published reports, unless noted otherwise.)

The classical nuclear family group consists of one adult male and one adult female, sometimes with one or more of their most recent offspring. The adult male and female can be said to be mated, to be linked to one another by a definite and personal pair bond. They associate with one another for long periods of time, perhaps for life in many cases. Both parents, the father as well as the mother, carry and take care of infants. This kind of small group seems to be the basic intraspecific social unit (not necessarily the only unit, but the one most frequently encountered) in all the populations of *Aotus, Callicebus moloch, Callicebus torquatus, Callimico goeldii,* and *Pithecia monacha* that I studied in Panama and Colombia. See also Appendix 1.

The large troop includes several adult males and females with a variety of young of different ages, possibly with occa-

sional visitors and friends. This sort of organization is characteristic of *Pithecia melanocephala* and several "*Cacajao*" type
forms (Mittermeier, personal communication, Le Nestour,
personal communication, Bates, op. cit., Durham, personal
communication, and Hernández-Camacho and Cooper), *Alouatta caraya* (Locker Pope, op. cit.), many or all of the denser
populations of *Alouatta villosa, Lagothrix,* the Common Spider
Monkey (at least *Ateles* "*geoffroyi*" and "*belzebuth*"), *Saimiri,*
and the capuchins, *Cebus capucinus* (Freese, op. cit.), *nigrivittatus* (Oppenheimer and Oppenheimer, op. cit.), *apella,* and
albifrons wherever they are abundant (but see also below.)

Troops of the same and different species differ in both
average and maximum sizes. Groups of more than a dozen individuals are common. There are rumors of enormous groups
of 100 or more individuals. Squirrel Monkeys may tend to have
the largest troops, or occur in large bands most frequently.
The sex ratio is supposed to be approximately equal in at least
one of the more gregarious species, *Alouatta caraya,* but adult
females usually outnumber adult males in such other forms
as *Alouatta villosa, Cebus capucinus,* and Common Spider
Monkeys. It may be assumed that all troops have a certain
amount of internal social structure. At least, many of the members must have special roles to play as leaders or followers,
initiators or imitators, adults or young, etc. The various roles,
and the individuals that fill them, are easy to recognize in
restricted conditions, when animals are confined in captivity
or concentrated around artificial food sources. They can be
more obscure in ceboid troops in the wild. The largest natural
troops sometimes give the superficial and misleading impression of being composed of similar and interchangeable particles
in a homogeneous social medium or continuous flux.

The appearance of simplicity is enhanced by several factors
that are not at all coincidental. Many New World primates
have weak dominance hierarchies, or establish and maintain
superior-inferior relations by subtle or inconspicuous interactions. Sexual relations within troops also tend to be variable

and intermittent, perhaps even promiscuous. Adult males of troop-living species seldom or never take intensive care of infants. Thus, the links between mothers and their offspring may be the only restricted bonds immediately visible to an outside observer.

Several kinds of groups are intermediate in size between typical small families and typical large troops.

One kind is exemplified by the groups of five to ten, or slightly more, individuals that are common among the *Saguinus fuscicollis* and *S. graellsi* of the Caquetá and Putumayo (again see Appendix 1). These appear to be "extended" families, to consist of a few pairs of adults, probably real mates, with their infants and juveniles. Some or all of the younger adults probably are the offspring or descendants of the oldest or original pair. Males show parental behavior and may carry infants. Other species of *Saguinus* (certainly the *midas* described by Thorington, 1968a) occur in groups of similar size. So do *Leontopithecus rosalia* (Coimbra-Filho and Mittermeier, 1973a), *Cebuella pygmaea,* and larger marmosets of the genus *Callithrix* (this is implied by the anecdotal accounts of travellers and explorers). There is evidence, from observations in the field or in the laboratory, that these animals have strong pair bonds and male parental behavior. It seems likely, therefore, that many of their groups also are extended families.

Saguinus geoffroyi might appear to be exceptional among its congeners. In central Panama now, most groups of the species seem to be small nuclear families. This may be a recent and somewhat artificial development. As noted above, the species has invaded new areas made available by human agricultural activities. It also is hunted intensively (not so much for food but in order to obtain young for sale as pets). Both factors would be expected to bring about a reduction in average size of groups. Perhaps the intrinsic social tendencies of *geoffroyi* would produce medium-sized groups like those of *fuscicollis, graellsi,* and *midas* in the absence of human interference. Conversely, a decrease or scattering of the populations of some other species of tamarins and marmosets might reduce them to smaller nuclear family units like those of present day *S. geoffroyi* and *Callimico*. The difference between nuclear and extended families may be quantitative rather than qualitative, little more than a reflection of greater or lesser crowding.

Other groups of intermediate size may derive from a different source. Some species that are found in large troops in some areas and situations also occur in smaller groups in other circumstances. The smaller assemblages may be composed of only one adult male plus several females and young. Such groups seem to be typical of the *Cebus capucinus* of Barro Colorado Island (Oppenheimer, op. cit.) and the *Alouatta seniculus* of the parts of the Meta and Caquetá that I visited. They could be called polygynous families. It might be more revealing to classify them as reduced troops. The adult males apparently do not usually carry the young. Their sexual relations are inconstant insofar as attention may be switched from female to female with considerable frequency. Groups of this kind may be characteristic of highly gregarious forms in suboptimal conditions, when populations are not abundant. Neville (in press) found groups of *seniculus* with more than one adult male in Venezuela where the species is not quite as rare as in the Meta and Caquetá. (Even in large troops, one or more adult males may secure preferential access to oestrous females. When and if so, the effective reproductive relations may be not very different from those in a reduced troop.)

Some students of Old World primates have contrasted single male troops with age-graded-male troops with multi-male troops. Doubtless the distinctions are real, but I do not think that this sort of classification is useful when applied to ceboids.

There can be other social but nonsexual or nonparental links between individuals both within and without the basic units, the family and the troop.

Species that are organized into nuclear families may exhibit traces of wider or larger intraspecific social relations in particular circumstances. Several family groups of Colombian

Pithecia monacha may congregate in the same large trees at night, or take shelter from heavy rains during the day. Night Monkeys may come together in large numbers in favored fruit trees in southeastern Peru (Durham). Neighboring pairs and families of *Callicebus moloch* may react differently to one another and to strangers, and perform communal defense patterns when populations are high (Mason, 1966).

Large troops, on the other hand, can split up into fractions, for longer or shorter periods, and still reassemble to re-form the original groups. The temporary subdivisions are various, including sexual partnerships ("consorts"), less obviously interested groups (friends), and special age and sex classes. The latter are most conspicuous in *Saimiri* (Thorington, 1967 and 1968b; see also Fairbanks, 1974, and Mason, 1974) and *Ateles* (comments in Eisenberg and Kuehn, op. cit.).

Common Spider Monkeys (at least "*geoffroyi*") would appear to occur in units of very different sizes more often than any other ceboid studied. The larger groups of "*geoffroyi*" and "*belzebuth*" also seem to split and reassemble with remarkable frequency. It has been suggested by Wagner (op. cit.) that much of this variability is a response to human hunting. Of course, the pressure of hunting upon spider monkeys is apt to be particularly strong. Wagner's suggestion is plausible. But the tendency to vary, to alter groupings in different conditions, almost certainly was selected for before hunting became very serious. Possibly it originated as an adaptation to the feeding habits of the monkeys themselves. For large animals that eat foods that are neither always abundant nor evenly distributed in space, it might well be advantageous to be able to make a voluntary choice between scattering and joining forces, depending upon the actual distribution of foods in any given area at any given time. (Reynolds and Reynolds, 1965, have suggested a similar explanation for the even more variable social arrangements of the larger and perhaps more nearly omnivorous chimpanzee, *Pan troglodytes*.)

All populations of all ceboids studied in the field have been

found to include a few single individuals apparently independent of groups. It is not known if their isolation is long-
sustained or not. My impression is that "loners" are least common in *Saimiri*.

At first glance, it might appear that there is almost every
conceivable intergradation among social organizations of New
World primates. Perhaps the proof of such a continuum will
be demonstrated by further studies. In the present state of our
knowledge, however, there seems to be a significant hiatus
between families, no matter how extended, and troops, no
matter how reduced. Even when groups are of the same size,
the two types may be distinguished by the nature of the internal sexual and parental relations within them. The visible
extremes may reflect a basic dichotomy.

The two types may be equally specialized in different ways.

It is noteworthy that family units are widespread among
ceboids. They occur in at least half of the living genera, and
proportionately even more of the species.

Another variable is spacing among individuals of the same
group. Some species are more consistently cohesive than others.
Of all the forms that have been studied in detail, close crowding among conspecifics seems to be most typical of *Saimiri*
and *Alouatta villosa*. (*Saimiri* may divide into subgroups, but
the divisions are not often far apart, and each is composed
of several individuals that usually stay within a few meters
or less of one another.) The cohesiveness of howlers and Squirrel Monkeys may help to explain some of the peculiarities of
their signal systems. See Chapter 5.

There are correlations between ecological features and social
organizations. Some examples have been cited, but others come
to mind. The usual slight gregariousness of *Aotus* may be
connected to nocturnality. Most other nocturnal animals of
other groups, certainly the overwhelming majority of those
that are both nocturnal and arboreal, are equally nongregarious. The close crowding of the vulnerable and opportunistic
Saimiri may be useful in defense or feeding.

Such simple correlations are not, however, as common as might have been supposed. A substantial number of the known cases involve minor rather than major aspects of social organizations, or interspecific rather than intraspecific relations (see below). It does, in fact, seem to be characteristic of ceboids that there is remarkably little general correspondence between the two basic types of intraspecific arrangements, families and troops, and broad preferences for either habitats or foods (Moynihan, 1973). There are both species that live in troops and species that live in families among the primarily or exclusively vegetarian forms. There also are both kinds of species among the animals that prefer insect foods. Percentages of highly and poorly gregarious species are much the same in many stages of succession from young scrub to mature forest in many areas. Densities of populations may also be irrelevant in this connection (if not for other facets of social behavior— again see below). Both *Callicebus moloch* and *Saimiri* usually are abundant wherever they occur. They are concentrated in different ways, but the average number of individuals per unit area per unit time may be high in both cases. Both *Aotus* and *Cebus albifrons* are dispersed on the average. The *albifrons* occur in troops, but the troops themselves are scattered.

It is quite possible that new and subtle correlations will eventually be discovered. (The spatial distribution and temporal patterning of food items are obvious subjects to be investigated in more detail.) Nevertheless, the facts cited above are real, and they are suggestive. Social behavior is not only an adaptation to external environments. On the available evidence, it would appear that different kinds of social organizations can work equally well in a variety of contexts, to permit or facilitate almost any kind of exploitation of environments within the (admittedly limited) range of niches occupied by diurnal ceboids at the present time. Part of the reason may be that many of these monkeys are more generalists than specialists. They can be flexible. Flexibility in some respects in the short run should favor stability in the long run in other respects. What may be supposed to have happened, on occasion, when a species or population was confronted with a difficulty in reconciling its social relations with its manner of obtaining food and other resources, is that it proved to be "easier" (more advantageous) to modify the methods of procurement than to change the social institutions. Simply because some or many of the resources used by ceboids can often be obtained and exploited by any one or all of several different strategies, some groups may have had the "choice" of adapting their societies to their economies or of adapting their economies to their societies. (The phraseology is anthropomorphic for the sake of brevity. It could be recast in terms of selection pressures.) Most other animals, nonprimates, do not have the same degree of freedom of choice. In a limited way, ceboids appear to have become partly independent of certain aspects of their surroundings. Perhaps there is a general trend in this direction among primates. Some ceboids seem to illustrate a moderately advanced stage in the process.

Sex and Aggression

One or two further points may be mentioned, largely for historical and comparative reasons. They concern sexual behavior, or more precisely, the relative frequencies of sexual and related acts, and their roles in social life apart from reproduction.

It has long been known, since the beginning of introspection or before, that there are connections between sex and hostile behavior (attack, escape, and combinations of the two). In monkeys, it is convenient to distinguish between short-term and long-term connections.

Ceboids, like most other primates, show temporal fluctuations in breeding behavior. Some forms, such as *Saguinus geoffroyi* and *Callicebus moloch*, have distinct breeding and nonbreeding seasons. Others, such as the *Alouatta villosa* of central Panama, are less obviously seasonal, although the numbers of births and young reared successfully may be greater or lesser at different times of the year. Direct sexual patterns, copulations and copulation attempts, fluctuate in accordance with breeding. So do many hostile performances. The particular performances differ from species to species. There may be a decrease of overt hostility between sexual partners when copulations are most frequent. There may be an increase of disputes among individuals, usually males, competing for partners, usually females. These are recurrent short-term changes.

Old World monkeys vary seasonally in the same sorts of ways as do New World forms. They show similar changes in hostile behavior at corresponding stages of the breeding cycle. But their short-term variations may be superimposed upon different long-term levels of activity. Some of the Old World monkeys, especially baboons (*Papio*) and many macaques (*Macaca*), perform what appear to be sexual acts comparatively very frequently. Real copulations and attempted copulations between adult males and females in breeding condition are often repeated at short intervals. Equally or more common, and more ambiguous, are patterns such as "soliciting" and "mounting" among both young and adults of the same or different sexes, in or out of the breeding season. These latter patterns are related to, or derived from, sexual behavior, but they are not reproductive and must have other functions. They have been called "pseudosexual."

Ceboids tend to be less demonstrative or more covert. Real copulations tend to be less frequent. Mountings by young or among individuals of the same sex are usually rare. Soliciting displays, apart from real attempts at copulation, are used sparingly or are quite absent in most species under natural conditions.

It was at one time thought that sexual attraction and the performance of sexual patterns were the main forces holding primate troops together (see, for instance, Zuckerman, 1932). Recent studies have demonstrated that additional factors are involved and may be more important. The data from ceboids would strongly support this conclusion.

Why should there be differences in frequencies of sexual and sexually derived acts among different kinds of monkeys? Part of the answer may lie in the nature of the interaction or interference between sex and aggression. Most ceboids seem to be less aggressive among themselves than are macaques and baboons. (Doubtless this is one of the reasons why their dominance relations are unobtrusive and appear to be comparatively relaxed or easygoing.)

Aggression must have both advantages and disadvantages. The obvious disadvantage is that it may lead to fighting, and fighting may lead to physical injury. (It also takes up time that could be profitably devoted to other pursuits, feeding or maintenance activities, or just plain resting.) The possible advantages are diverse and, as usual, different in different species. Some of them are discussed in K. Lorenz (1966). Many students of behavior, Lorenz among others, have suggested that the aggressiveness of macaques and baboons is correlated with the fact that they are terrestrial or semiterrestrial and may stray into open country. They are very exposed to predators, and they must occasionally find themselves far from escape routes to the nearest trees. When and if an individual or group of macaques or baboons should be trapped in the open, the only possible defense may be attack. A determined attack is always at least startling. The effort should be particularly worthwhile for macaques and baboons because they are among the largest and most imposing of monkeys. This should encourage them to be aggressive during interspecific encounters. The more aggressive individuals and groups may be likely to attack more successfully, and, therefore, leave more descendants, than those that are less aggressive. The extension of increased aggression to other spheres of activity may be inevitable. Intraspecific and interspecific reactions are quite different in some respects, but there is almost always some "seepage" between them. Qualities that have become hypertrophied in one context are apt to appear in others. Thus, one might expect that selection in favor of aggression against predators might also tend to produce increased aggressiveness in other situations, including intraspecific encounters.

The performance of sexual or sexually derived acts is a way of controlling aggression among conspecifics. However much the onset of sexual behavior may stimulate such conflicts as disputes among rivals, the movements of overt attack and copulation are physically incompatible in the end or in their extreme forms. They cannot be carried to a climax absolutely simultaneously. Thus, the release of sexual behavior or sexual feelings by the performance of soliciting patterns to induce mounting may impede or even cut off attack. The various sexually derived patterns of such monkeys as macaques and baboons are, in fact, used by subordinate individuals to denote submission or promote appeasement. Their proliferation and elaboration would seem to have been selected for to subserve these purposes. See also Wickler (1967) and the comments and references in A. Jolly (1972).

Most New World primates may be less aggressive among themselves than are macaques and baboons because it would not be adaptive for them to be as aggressive toward potential predators. Being arboreal, they may be less likely to find themselves without avenues of escape. Being comparatively small on the whole (partly in correlation with arboreality), they may have less chance of attacking predators successfully. It seems probable that their sexual behavior is relatively simple because they do not need a very complex array of devices to control aggression.

There are a few exceptions that may prove the general rule. Adults of *Callicebus moloch* copulate more frequently, during their breeding seasons, than do any other of the better known ceboids. They also are particularly aggressive. As noted above, much of the territorial defense of the species is by means of vocalizations; but physical combats are also relatively, if not actually, common among neighboring territory owners, at least at Barbascal in the Meta. (Kinzey, 1972, seems to have misinterpreted comments in earlier papers such as Moynihan, 1966.

It is true that *moloch* is less aggressive than the nocturnal *Aotus*, but it engages in serious fights more frequently than all or most of its other relatives.) Selection may have favored the development of a moderately high degree of aggression among *moloch* individuals for several reasons. The territories of the species are comparatively small, which may facilitate trespassing, and, therefore, increase the need for active repulsion. Most individuals of the species may have time to spare because their food is superabundant. The teeth of both males and females are small or dull enough to reduce the risk of mortal injury. And the predation factor may also be significant in an unexpected way. Again as noted above, *moloch* seems to be very well protected against predators. This could have curious side effects in certain circumstances. One of the supplemental disadvantages of overt fighting is that it enhances the conspicuousness of the fighters and may attract the attention of dangerous passersby. In the case of baboons and macaques, this disadvantage must be accepted for the sake of the accompanying compensation. Aggressive baboons and macaques may be more conspicuous than would be more timid animals, but they can also defend themselves more effectively. In the case of *Callicebus moloch*, the disadvantage must be less because conspicuousness is not likely to increase a danger that is usually minimal.

Male *Saimiri* perform a rather obscene "penile erection" display before copulations (Ploog and Maclean, 1963). They may use the same display, presumably as a derived pattern, as threat or greeting in a variety of other situations. It is interesting, therefore, that individuals of the species show a relatively large amount of hostility toward one another, not so much actual fighting as brief chases and retreats and other ritualized and unritualized patterns containing components of attack and escape. Their hostility toward one another could be a by-product of their frequent close crowding. Crowding is always irritating.

The penile erection of *Saimiri* may be the only ceboid display that is both derived from a copulatory movement and used regularly during nonsexual encounters. Other possible examples are dubious or enigmatic. Thus, individuals of *Cebuella* and species of larger marmosets sometimes perform so-called "genital presentation" in hostile circumstances in the laboratory (Epple, 1967; Epple-Hösbacher, 1967). These performances might be of sexual origin; but nothing is known of their functions or effects in the wild (I did not see them among the Pygmy Marmosets of the Putumayo), and it is also conceivable that they have been partly or wholly derived from territorial scent marking (see Le Roux, 1967). The great majority of the sexual patterns of most American monkeys certainly have purely sexual objectives, even if they are occasionally misdirected or happen to procure other than sexual benefits.

Frequencies of aggressive and sexual or related acts appear to have varied, increased or decreased, together during the history of the primates. The linkage between them seems to have been persistent but at one remove. Sex and aggression have been coupled in the long run because they are often difficult or impossible to combine at any given moment. This is not a paradox. One kind of behavior can suppress or modify the expression of the other. The evolution of the antidote has conformed to that of the poison.

Interspecific Behavior

All New World monkeys must cope with other species in addition to predators. Some respond overtly to competitors and encourage or tolerate associates, while others do not. The range of possible interspecific reactions, and their apparent absence in particular cases, may be illustrated by brief accounts of the situations in several different regions.

In Panama. There are six species and seven distinct kinds of ceboids in Panama: the tamarin *Saguinus geoffroyi*, the Night Monkey *Aotus*, the howler *Alouatta villosa*, the Squirrel Monkey *Saimiri*, the capuchin *Cebus capucinus*, and two spider monkeys of the genus *Ateles*—a large black *"fusciceps"* and a smaller red *"geoffroyi."*

The Night Monkey occurs throughout most of the country, but is segregated from the other local primates by its nocturnal habits. Its presence may still have an effect upon other monkeys. It eats some of the same fruits and probably insects as the others. Its size alone would suggest that it is most likely to take foods that might otherwise be eaten by *Saguinus, Saimiri,* or *Cebus*. It may compete more strongly with other nocturnal animals, especially some of the arboreal and semiomnivorous raccoonlike carnivores, such as the Kinkajou *(Potos flavus)*, cacomistles *(Bassariscus)*, and olingos *(Bassaricyon)*, although I have never seen any personal interactions among them. This is not really surprising: complex interspecific behavioral responses seem to be much rarer among nocturnal species than among diurnal ones, which might suggest, in turn, that such reactions usually are strongly dependent upon visual cues.

Alouatta villosa also is widespread. It certainly is, or was, in actual or potential contact with all the other diurnal monkeys of Panama, but it does not seem to have developed any very distinctive or unusual overt behavioral responses to them. Groups of howlers frequently find themselves in the same trees as capuchins and red spider monkeys, perhaps more often than would be inferred from the descriptions of Carpenter (1934 and 1935). They seem to be drawn by some of the same fruits or other features of the physical environment as the other two species. When encounters do occur, the howlers are definitely subordinate. They may be larger or heavier than the other forms but they also are less active and excitable. Capuchins often "tease" or threaten howlers, making inhibited or intentional movements of attack (personal observation and Oppenheimer, 1968). Although quite mild, these antagonistic patterns are sometimes effective. The arrival of a group of capuchins in a tree in which howlers are already established may induce the latter to move on rather earlier than they would have done otherwise. The relations between howlers and red spider monkeys are even less friendly. Richard (1970) says that groups of howlers retreat before groups of spider monkeys in the forest on Barro Colorado Island. I saw several furious disputes with much high-intensity threat between adult male howlers and an adult male spider monkey in the same forest. Once the male spider monkey actually drove a male howler out of a tree and down to the ground. This is the only time I have ever seen a fully adult male *villosa* on the ground in the wild. Although even this dispute was not accompanied by physical combat, actual wrestling, or biting, it certainly placed the howler in an awkward and vulnerable position. (Spider monkeys are a recent reintroduction to Barro Colorado, but the individuals involved in these incidents had been living free for some years and had adjusted well, apparently normally in many respects, to the local environment.)

Thus, it would seem that both capuchins and spider monkeys may have deleterious effects upon howlers in central Panama, at least in the short run.

Howlers take some fruits that would otherwise be available to capuchins and spider monkeys, but they also eat many leaves that the other forms do not. One might have expected, therefore, that competition between the howlers and the others would be unimportant. The overt hostility of the spider monkeys may be evidence to the contrary. Since disputes are dangerous and time-consuming, natural selection would have favored interspecific hostility, as such, only if it could provide some significant reward, however small. Possibly the aggressiveness of spider monkeys toward howlers is an extrapolation of aggressiveness among themselves. It is perhaps more probable that howlers and spider monkeys compete for food more seriously than the currently available data would suggest during unusual but critical periods of scarcity or need to use one another's living space, travelling routes, or resting sites.

Offhand, one might also have expected howlers to react

strongly to the leaf-eating tree sloths, *Bradypus* and *Choloepus*, and perhaps been equally mistaken. Visible responses between monkeys and sloths are virtually nil, perhaps because the sloths themselves are so nearly completely nonreactive.

The two forms of spider monkeys in Panama replace one another geographically. The black form occurs in the eastern part of the country, the Darien. The red form used to range through most of the center and west. Unfortunately, the red animals have been hunted so intensively that they have disappeared from many areas, including most of the Canal Zone and the neighborhoods of the cities of Panama and Colón, the precise areas where they approached the range of the black form. It is no longer possible to observe reactions between the two forms in the wild. Until a few years ago, however, individuals of obviously intermediate size and color used to appear for sale as pets in the markets of Panama City from time to time. They varied considerably: some were almost exactly half way between the two extremes; others were more like typical "*fusciceps*"; and still others were more like typical "*geoffroyi*." This would indicate that introgression must have occurred. Probably the peculiar specimens obtained by some early mammal collectors, which have been called *Ateles* "*rufiventris*" and "*geoffroyi grisescens*," were also intermediates, first and later generation hybrids exhibiting different mixtures of parental characters (see Hershkovitz, 1949, and Hill, 1962).

The fact that "*fusciceps*" and "*geoffroyi*" interbred is somewhat remarkable, as the size difference between the two forms is substantial, and male and female spider monkeys can be hostile to one another. Copulations among spider monkeys are unlikely to occur without more or less prolonged "courtship" beforehand. This would suggest that "*fusciceps*" and "*geoffroyi*" or their hybrids once associated in mixed troops in the border region.

The distribution of color types may throw additional light on the interactions between spider and howler monkeys. It may be significant that the spider monkeys of most of Central

America are reddish and those of the Amazon basin largely black, while the howlers of Central America are blackish and those of the Amazon bright red. The transitions of changes from red to black and black to red do not coincide exactly, but they are only a few hundred miles apart. They could be examples of "character displacement" (Brown and Wilson, 1956), selected expressly to enhance the apparent differences between overlapping or adjoining forms. These displacements could have several functions: they may be relevant to relations between forms within a genus, between the two types of spider monkeys and between the two types of howlers; they may also affect relations between genera. Considering the dominance involved, it is conceivable that one of the functions of the howler change is to reduce or restrict aggression from spider monkeys. It is a general rule that, all other factors being equal, animals are likelier to attack animals that look like themselves than those that look very different. This is another example of the occasional correlation of intra- and interspecific behavior. Howlers that differ from spider monkeys in color may be less likely to be attacked than are howlers that are similar in color. If this suggestion is correct, it implies that howlers have adapted to spider monkeys rather than vice versa, or, in other words, that it is the color of the local spider monkeys that has determined the color of the howlers in most of these areas.

Spider monkeys may be the most important primate competitors of howlers, but the reverse is not necessarily also true. In Panama, *Cebus capucinus* seems to be at least as important as the local form of *Alouatta* as a competitor of *Ateles* "*geoffroyi.*" These capuchins and spider monkeys eat more of the same foods. Interestingly enough, the social relations between them are quite different from those of either with howlers. They are much more consistently close and may include more friendly components. They have been studied in detail only on Barro Colorado Island. Here the reintroduced red spider monkeys showed a marked tendency to join bands of *Cebus* as soon as they were let loose. They have maintained this ten-

dency ever since. At the present time, when individuals of the
two species are associating, it usually is difficult to decide who
is leading or following whom. Either species may be found in
the forefront of a mixed group (A. Richard, personal communication). There are occasional violent quarrels between the
species, but I should not be prepared to say that they are relatively more common than intraspecific fights among either capuchins or spider monkeys alone. Then, too, the two species
sometimes stay together for hours on end with every appearance of amiability.

The social integration of these spider monkeys and capuchins may be facilitated by the similarity in form of many of
their signals, facial expressions, and vocalizations. Some of the
resemblances probably are due to inheritance from a common
ancestor, while others may be examples of social mimicry (see
below and also Moynihan, 1968).

The local tamarin, *Saguinus geoffroyi*, is involved in a rather
different cluster of relations. It does not seem to interact with
spider monkeys or howlers to any appreciable extent. It is
largely separated from them by its preference for dense, often
second growth, tangles and scrub or low forest. Also, it is too
small and too partial to insects to be either a serious competitor
or useful associate even in intermediate or mixed vegetation. It
may impinge upon the sphere of the local capuchins slightly
more frequently. The two species may take some of the same
kinds of insects, but, again, they usually prefer different habitats. When and if they do manage to approach one another,
the tamarins usually retreat after a few minutes, sometimes
precipitously.

Competition between the tamarins and the not very large
Squirrel Monkeys is, or could be, more serious. The two species
are either segregated geographically or they overlap only in
areas where the populations of both are sparse (see references
in Moynihan, 1970b). This would suggest that the separation
itself might be the result of "competitive exclusion" (Hardin,
1960).

Relations between the tamarins and birds seem to be more subtle and complex. Flycatchers of the family Tyrannidae occur throughout Panama. Many of them are abundant; all are diurnal. Some of them eat many of the same insects, berries, and other small fruits as the tamarins do, and in the same areas. (The feeding techniques of the monkeys and birds are different, but many items of food are accessible to both.) Some of the sounds uttered by certain species of flycatchers, e.g., *Myiozetetes* spp., *Tyrannulus elatus, Legatus leucophaeus, Elaenia flavogaster*, and *Megarynchus pitangua*, are similar to (perhaps even indistinguishable from) some of the vocalizations of tamarins. The latter may recognize potentially favorable areas with lots of suitable food simply by the large numbers of flycatcher sounds coming from them.

Some of these resemblances among the calls of different species might also be due to social mimicry. The mechanics of the adaptations that would be involved are discussed in more detail in Moynihan (1968 and 1970b).

One conspicuous difference between tamarins and flycatchers is timing. All the flycatchers become active at dawn and feed most intensively in the next few hours, while the tamarins (as noted above) may not start to move until well after dawn and continue to feed fairly steadily throughout most of the day. The tardiness of the tamarins, which is very unlike most other diurnal mammals as well as birds, may be a direct adaptation to reduce the strength and frequency of interspecific competition. If so, the partial segregation of flycatchers and tamarins would be strictly comparable to, although less extreme than, the separation between tamarins and Night Monkeys.

Double-toothed Kites (*Harpagus bidentatus*) often accompany bands of capuchins. They may also join and follow some of the larger groups of tamarins in areas where capuchins are absent. They appear to be feeding on small animals, insects and arboreal reptiles, disturbed by the monkeys. This sort of relationship is more parasitic than commensal. The kites contribute little or nothing to the monkeys in return for services rendered. Unlike some other associates of other species (see below), they do not even act as sentinels. They do not seem to notice predators or utter warning calls any earlier than the monkeys themselves. The tamarins seem to ignore the kites completely. This may be advantageous, as it should save time and worry. The tamarins are neither large nor agile enough to drive away the kites. It is remarkable, nevertheless, because *Harpagus* has much the same general appearance or *gestalt* as other birds of prey that release panic. The original development of indifference to the kites may have been difficult and dangerous.

The scrub and forest areas of the Pacific coast of Panama are inhabited by several kinds of diurnal arboreal squirrels which are much the same size as tamarins and eat some of the same vegetable foods. The only species I have observed at any length in the central part of the country is *Sciurus granatensis*. Individuals of this species often visit the same places as tamarins, sometimes repeatedly on the same days, but only very rarely are the two species present at any given site absolutely simultaneously. This does not seem to be due solely to differential timing of the kind that separates tamarins from Night Monkeys or flycatchers. The squirrels are active when the tamarins are. There seems to be another kind of avoidance mechanism involved, a combination of temporal and spatial arrangements. On some of the very few occasions when I did see individuals of the two species fairly close together, the squirrels showed obvious signs of hostility, apparently predominantly alarm. Perhaps the squirrels simply tend to refrain from approaching tamarins whenever they know that the latter are around.

The interspecific reactions of Squirrel Monkeys in western Panama are much less well known. The only other diurnal primate that is still common in the parts of the region where I have observed *Saimiri* is *Cebus capucinus*. The relations between the two may be similar to those between tamarins and *Sciurus granatensis* in the rest of the country. Individuals of the two species may come near to one another occasionally, but seldom or never stay together for more than a few minutes. I did not see any unmistakable overt communication between them during my own brief observations. The Squirrel Monkeys sometimes uttered alarm notes when capuchins were not far away, but it was difficult to be sure of the releasing stimuli. A native hunter on the Burica Peninsula said that Squirrel Monkeys often appear to be frightened of *capucinus*, and that he once saw one of these capuchins chase and actually bite a Squirrel Monkey. Thus, it seems that associations between the two species in this region are rare and brief because the Squirrel Monkeys try to avoid or retreat from the larger animals.

Whatever they may be, the relations between *Saimiri* and *Cebus capucinus* in Panama cannot be similar to those between *Saimiri* and *Cebus apella* in parts of South America (see below).

Like *Saguinus geoffroyi*, the Central American form of *Saimiri* overlaps squirrels, but with exactly opposite social results. Near Puerto Armuelles, I saw many squirrels of two species, *Sciurus variegatoides* as well as *granatensis*, in an area inhabited by both Squirrel Monkeys and capuchins. Both species of squirrels showed a strong tendency to join and follow the Squirrel Monkeys. This is particularly interesting if, as suggested above, the Central American Squirrel Monkeys are somewhat less insectivorous than the local tamarins. The largely vegetarian squirrels may compete with their associate *Saimiri* for food relatively more often or strongly than with their nonassociate *Saguinus*.

The social preferences of the squirrels of the Puerto Armuelles area seems to be both one-sided and quite specific. The Squirrel Monkeys usually ignore the squirrels. The squirrels do not show any particular tendency to join or follow the local capuchins. Individuals of different species of squirrels do not tend to associate with one another.

The hunter who reported hostile reactions between Squirrel Monkeys and *capucinus* also said that groups of the former may be accompanied by White Hawks (*Leucopternis albicol-*

lis). These birds of prey also feed on reptiles and large insects (Wetmore, 1965). They may associate with Squirrel Monkeys for the same reasons that Double-toothed Kites associate with capuchins and tamarins.

In the Upper Amazonian Region of Colombia. The species I saw in the Caquetá were the tamarin *Saguinus fuscicollis*, the howler *Alouatta seniculus*, the saki *Pithecia monacha*, the titi *Callicebus moloch*, the Woolly Monkey *Lagothrix*, the spider monkey *Ateles "belzebuth,"* the Squirrel Monkey *Saimiri*, and the capuchins *Cebus apella* and *albifrons*.

In the Putumayo, I observed some of the same species, plus the tamarins *Callimico goeldii* and *Saguinus graellsi*, the Pygmy Marmoset *Cebuella*, and the titi *Callicebus torquatus*. Other species must occur in the same areas (e.g., *Aotus*) or not very far away (e.g., *Pithecia melanocephala*).

In the Meta, the only species seen were *Alouatta seniculus, Callicebus moloch, Saimiri,* and *Cebus apella*. With the probable addition of the Night Monkey, this may be the complete list of ceboids surviving in the area visited (San Martín-Barbascal). The list has been reduced by human activities; hunting pressure is supposed to have been intense at one time. The original fauna of the area certainly included *Lagothrix* and *Ateles* and perhaps some other monkeys. (It should be stressed, however, that human interference can hardly explain the absence of *Saguinus fuscicollis*, for second growth remains, and hunters would not have exterminated tamarins while leaving larger forms. Even *Cebus* and *Saimiri* are more succulent than *Saguinus*.)

The primates of the Caquetá and Putumayo seem to pay little overt attention to small birds in most circumstances. Flycatchers are rather scarce in the areas where tamarins have been studied. They are more abundant in some of the habitats of *Cebuella*, but the marmosets merely dodge them when they fly by. (The nearest thing to a friendly bird-monkey relationship observed in the Amazon involved another kind of bird.

A flock of a widespread species of tropical jay, *Cyanocorax violaceus*, was found in a small and isolated patch of rather degraded wood near Rumiyaco. The fauna of this wood was impoverished, but it still included a couple of pairs of *Callicebus torquatus* and two small groups of *Saguinus fuscicollis* in addition to the jays. The three species reacted to one another's alarm calls and performed mobbing together. It was my impression that the jays also tended to stay with the tamarins when not disturbed. Possibly they had developed some sort of personal bond, perhaps encouraged by the lack of other animals to react to in this depauperate environment. Jays are among the more intelligent birds—see Chapter 7—and therefore presumably easily bored and predisposed to strike up new relationships. Other individuals of the same species do not seem to associate with monkeys in more extensive or richer vegetation.)

There is a more remarkable and regular arrangement between *Cebuella* and a species of pygmy squirrel (probably *Microsciurus napi* or *M. flaviventris napi*, Hernández-Camacho, personal communication). Squirrels of this species may prefer deep or old forest, but they also occur in second growth hedges. In this habitat, they are less common than the marmosets, but probably not actually rare. They are incredibly similar to the marmosets in appearance at a distance, almost exactly the same size and color, and not very different in shape. I know that the home ranges or territories of individuals of the two species are sometimes broadly overlapping, but I never saw any face-to-face or even moderately close contacts between them. They may be kept apart from one another by some short-term avoidance mechanism. There is supporting evidence from another site near Rumiyaco. Here I found a large tree, *Inga* sp., which split into four equal trunks a few feet from its base. Three of the trunks were covered with the marks of gnawing by pygmy squirrels. I saw one squirrel make some of the marks, while it was apparently stripping off and eating bark. Two of these trunks had no Pygmy Marmoset feeding holes in them. The third had a few holes on the lower part. The fourth trunk was absolutely riddled with marmoset holes and had only a very few squirrel marks. Obviously the two species were dividing the tree between them. The number of marmoset holes was about as great as usual in a feeding tree of the species. This might suggest that the marmosets had taken what they wanted and left the rest.

I have no other information on the diet of pygmy squirrels. They must be competing with the marmosets in one way, if no other. Neither the drilling of the marmosets nor the stripping of the squirrels can be good for the trees. The activities of either species must tend to have unfavorable effects upon the other, insofar as they may reduce the vigor or impair the chances of survival of an essential or useful source of food.

The social segregation of the two species, which may permit or facilitate their coexistence in the same places, seems to be partly a matter of timing. Indians of the Putumayo claim, and my own observations would tend to confirm, that Pygmy Marmosets are most active during the morning and late afternoon, and usually take shelter during rain. Pygmy squirrels, on the other hand, are active during the middle of the day and continue feeding during at least moderately heavy downpours.

There are larger squirrels of the genus *Sciurus* in the Amazon. The species *granatensis* extends into the Caquetá. Individuals of the local population(s) of the species do not seem to be concerned with or bothered by sympatric monkeys in spite of the fact that they take some *Jessenia* and probably other palm nuts. Quite in contrast, individuals of another, closely related species, *Sciurus igniventris*, are often found in association with *Saguinus graellsi* near Santa Rosa in the Putumayo. It is usually evident that the squirrels are following the tamarins.

Some monkeys of the upper Amazon have little chance to interact with other monkeys of other species because of partial ecological isolation. This is true of the *Cebuella* of the smallest and most isolated hedges and many *Callicebus moloch* of the most waterlogged swamps.

In other areas and environments, where there are eminent opportunities for interactions, particular species may still tend to ignore one another. *Saguinus fuscicollis* and *Pithecia monacha* provide a good example. Individuals of the two species often occur within sight or sound of one another in such areas as the vicinity of Valparaíso in the Caquetá, where their preferred habitats intergrade or intermingle in mosaic fashion. Their territories certainly can be overlapping, in this case literally as well as figuratively, as the sakis tend to remain higher than the tamarins. Partly because of this height difference, contacts between the two forms are seldom or never very close; probably they are essentially "random" or "accidental." Individuals of the two species apparently do not try to join up, but neither do they retreat from one another as long as some reasonable individual distance is maintained.

Much the same can be said of the reactions of both forms with *Saimiri, Cebus apella*, and *C. albifrons*. The tamarins and sakis neither follow nor avoid Squirrel Monkeys and capuchins. The lack of interest seems to be reciprocated, except in very unusual circumstances—see below.

It is suggestive, in this context, that competition between sakis and tamarins must be minimal. And probably neither *Saguinus fuscicollis* nor *Pithecia monacha* competes with *Saimiri* or *Cebus* spp. very strongly.

Competition between the Woolly Monkey and the Common Spider Monkey, on the other hand, must be rather intense in some ways, because the two species take some of the same kinds of fruits in the same places. But my own observations, information from local settlers and hunters in the Caquetá, and studies by other biologists (Lehmann, personal communication, Hernández-Camacho and Cooper, op. cit.), all indicate that they seldom or never occur in the same places at the same times, at least in Amazonian Colombia—see also below.

In these circumstances, the separation must be due to positive avoidance. The result is reminiscent of the *Saguinus geoffroyi–Sciurus granatensis* and *Saimiri–Cebus capucinus* relationships in Panama, although it may be achieved by less one-sided reactions. Woolly Monkeys and spider monkeys are similar in size, with comparable offensive and defensive weapons. They may, therefore, avoid one another mutually. A few observations would suggest that *Pithecia monacha* and *Callicebus torquatus* may be kept apart by a similar avoidance mechanism.

There are friendly reactions among some of the same and other species. They also vary in intensity. With its preference for tangles and certain kinds of thickets, *Saguinus fuscicollis* sometimes ranges along the interfaces between different types of forests, and it is one of the few species that may verge upon the most extreme habitats of *Callicebus moloch* in some areas— in the Caquetá if not in the Meta. When the two species are adjacent, the relations between them can be slightly complicated. The titis seem to pay little or no attention to the tamarins. The tamarins keep away from the titis most of the time. Yet they obviously are excited by some of the characteristic *moloch* vocalizations, especially the very elaborate, loud, and prolonged "Songs" which contain many purely and partly hostile components (Moynihan, 1966). They may start to move in the direction of a singer. Possibly they are drawn by sheer curiosity to see what the fuss is all about. The approaches seldom lead to anything very definite; the tamarins simply hang around for a while and then drift away again, presumably with their curiosity satisfied and having learned something new about their surroundings in the process.

The *fuscicollis* of the Putumayo also appear to be attracted by the similar "Songs" of *Callicebus torquatus*. Approaches to the latter may be encouraged by two additional features: both *torquatus* and *fuscicollis* are conspicuously black and rufous; the preferred habitat of *torquatus* usually is less sharply marked off from that of *fuscicollis* than are some of the types of vegetation inhabited by *moloch*. Thus, it is conceivable that *fus-*

cicollis and *torquatus* have developed specialized mutual links, viz., the joint mobbing cited above. Nevertheless, the links can be no more than occasional and intermittent, as the two species certainly spend more time apart than together.

Callicebus moloch and *C. torquatus* themselves may be able to hear one another's "Songs" (usually faintly in the distance) in some areas. I did not notice any mutual interference or co-operation, either "chorusing" or "jamming."

The closest interspecific relations involve *Saimiri* and *Cebus apella*. They show a maximum of interspecific gregariousness. Wherever they both survive, they apparently always occur together in large mixed groups of considerable complexity.

Squirrel Monkeys outnumbered capuchins in all the groups seen. A range of 2–10 capuchins and 20–50 Squirrel Monkeys per group seems to be usual in the Caquetá. This appears to be an accurate reflection of the real abundance of the two species, i.e., almost all individuals of both species are involved in the association.

The mixed groups of the Caquetá are very cohesive and probably enduring. They maintain themselves without much obviously organized leadership. Either species can, and frequently does, join and follow the other. There is some tendency for individuals of the same species to stay closer to one another on the average than to individuals of the other species, but exceptions are frequent. Individuals of either species can approach within a few feet, perhaps inches, of individuals of the other without provoking active hostility, either aggression or retreat. I have never seen the two species fight over a particular item of food, even though they may take some of the same kinds of insects and fruits at different times.

Thorington's account (1968b) of a small number of individuals of both species in a small forest in the Meta would suggest that the association between them may be less close, and more often hostile, than between individuals of the same species in the Caquetá. It is still persistent. The same individuals joined one another again and again during Thorington's study, and also during my own much briefer observations in the same area.

The attraction between *Saimiri* and *Cebus apella* may be reciprocal, and essentially similar to the bond between *Cebus capucinus* and *Ateles* "*geoffroyi*" on Barro Colorado Island. Probably the interspecific preferences of the Amazonian forms are strong. Certainly Colombian Squirrel Monkeys will resort to unusual expedients to avoid isolation. G. Fuerbringer (personal communication) saw a Squirrel Monkey following a group of *Saguinus fuscicollis* in very disturbed and impoverished scrub near Rumiyao. Mason (personal communication) saw a Squirrel Monkey with a group of *Callicebus moloch* in a small and isolated patch of wood in the Meta (not the same forest visited by Thorington and me). *Cebus apella* were absent from both sites at the time, and the two Squirrel Monkeys were the only members of their species present. (The Rumiyaco individual may have been a stray; the Meta animal may have been either a stray or the last survivor of a population destroyed by hunting or man-made alterations of the environment.) Presumably the two Squirrel Monkeys were so desperate for companionship that they were willing to put up with very unsuitable partners. *Saimiri* individuals do not associate with either *Saguinus fuscicollis* or *Callicebus moloch* in more favorable habitats where alternatives are available.

The specialized and sometimes exclusive nature of the *Saimiri-Cebus apella* relationship is revealed by comparison with *Cebus albifrons*, which is potentially capable of impinging upon the other two forms socially and ecologically, as well as geographically. Many of its intraspecific behavior patterns, including ritualized displays, are very similar to those of *apella*. Some social signals of both capuchins are not very different from those of *Saimiri*. The ranges of foods taken by various populations of *albifrons* may be more or less intermediate between the equivalent ranges of *Saimiri* and *apella*, which also, of course, overlap one another. It would be expected, therefore, that *albifrons* individuals might tend to find themselves in ambiguous or difficult situations. They have solved their problems quite neatly, but by diverse methods. In the Caquetá, where their diet may be most *Saimiri*like—with a preponderance of soft fruits—and where *apella* is common, they are socially isolated. The individuals of *albifrons* I saw near Valparaíso were not associated with other monkeys. Local settlers agree that the species does not usually occur with either *Saimiri* or *apella* in this region, except, very temporarily and occasionally, in fruiting trees where food may be superabundant. Conditions are very different in the Putumayo near Santa Rosa. There *apella* is absent. The resident *albifrons* eat *apella*like foods, hard nuts, and also associate with *Saimiri* in exactly the same ways, with the same frequency, and in the same proportions, as *apella* elsewhere. The Santa Rosa relationship might be ideal for *albifrons*. Its drawback is that it is nonfeasible in other areas where *apella* prevails.

Other substitutions may occur in special circumstances.

Cebus apella has been seen with *Ateles* "*belzebuth*" at the southern end of the Macarena mountains (Hernández-Camacho and Cooper). Presumably *Saimiri* was absent.

Callimico goeldii may be an "outrider" of mixed groups. On the island in the Rio Guineo near Puerto Umbría, where *Saimiri* does not live now, pairs and families of *Callimico* sometimes associate with groups of *Saguinus fuscicollis*. Some of the *Callimico* that I observed were following *fuscicollis* directly or along a parallel course to the side. On the adjacent mainland, where *Saimiri* survives in some numbers, however, the local Indians and settlers say that the *Callimico* always (sic!) associates with Squirrel Monkeys, presumably instead of *fuscicollis*. Possibly the island *Callimico* join *fuscicollis*, like the stray *Saimiri* cited above, simply because they have no other choice. *Cebus apella* is rare around Puerto Umbría, and *Cebus albifrons* is no longer common. I did not see either capuchin in this area. None of my informants could tell me anything about possible interactions between *Callimico* and *Cebus*. In view of the size differences involved, it might be supposed

that the tamarins would tend to avoid the capuchins. This is
not a foregone conclusion—as indicated by observations of
other individuals and species.

In Other Parts of South America. Mittermeier (personal communication) found small sakis, presumably *Pithecia cf. "monacha,"* with other monkeys such as *P. calva (sensu stricto)* in
parts of Amazonian Brazil. Perhaps the local sakis behave
somewhat differently from those of Colombia. Mixed groups
of blackish *monacha* and white *calva* must be striking, almost
like an advertisement for scotch whiskey.

According to Husson (op. cit.), *Cebus apella, Saimiri,* and
Saguinus midas often occur together in Surinam.

Sanderson (op. cit.) claims to have seen feeding aggrega-
tions of tamarins, howlers, Woolly Monkeys, spider monkeys,
and capuchins, probably also in the Guianas. This record would
be highly suggestive if it were reliable (see below), but the
published account is unfortunately brief and vague.

Comment. The differences among the interspecific reactions of
different species are hardly less great than among their intraspecific responses. Again, significant correlations are not always easy to identify or interpret.

There is some direct correspondence or resemblance between
interspecific and intraspecific social units, but it is only partial,
and seems to hold good only for extreme types. Some of the
species that are least gregarious among themselves (*Aotus,
Callicebus moloch*) seldom or never occur in mixed groups,
and Squirrel Monkeys, which are extremely gregarious among
themselves, occur in mixed groups very frequently, but there
are many anomalies among the other species. Most howlers do
not establish long sustained or friendly associations with individuals of other species in spite of the fact that they are very
gregarious among themselves. *Pithecia monacha, Cebus,* and
Saguinus include populations or forms that appear to have

similar intraspecific reactions but very different interspecific relations.

There is another partial correspondence, in this case probably indirect, between the development of interspecific organizations and feeding habits among diurnal species. The monkeys that are least likely to mingle with other species are all largely vegetarian. *Alouatta villosa* agrees with *Callicebus moloch* in this respect. Conversely, the establishment of friendly interspecific bonds is characteristic of many, although certainly not all, of the more nearly omnivorous or preferentially insectivorous forms. Unfortunately, the simplicity of this dichotomy is spoiled by more exceptions, most notably the Common Spider Monkey, which is frugivorous but still associates with other species.

Perhaps some of the apparent discrepancies can be resolved by a consideration of selection pressures. Like aggression, the habits of either mixing or not mixing probably have both advantages and disadvantages. An individual of one species associating with individuals of another species may find more food and other resources, or find them more easily, than it would if it were by itself alone. It may be able to share or appropriate some of the good things discovered by its companions. It must also risk greater competition, as its companions may take some of the food it has discovered for itself, or may occupy the most favorable resting places and lookout sites. The addition of individuals of other species to a group may make all the animals more conspicuous, and, therefore, attract the attention of predators, although the extra eyes and ears may also facilitate early detection of dangers, and increase the chances of either successful escape or concealment.

As far as I can tell, the disadvantages of mixing should be much the same for all ceboids (of comparable sizes, in similar areas, etc.). This would indicate that the advantages must be greater for those that do mix than for those that do not. Companions may be helpful to the more insectivorous species because they flush potential prey. Still, one does not see flushed prey caught very frequently, by the monkeys themselves, in the wild. The most highly insectivorous types do not necessarily form heterospecific groups more frequently than do species that are only moderately insectivorous (compare *Saguinus fusciollis* with *Saimiri* and *Cebus apella*). Flushing is irrelevant to spider monkeys. All this would seem to suggest that either the antipredator factor or the search for resources other than insects or both is determinant for some or all of the species of mixed groups.

It remains to explain why the addition of extra eyes and ears to detect extra resources or predators should be more advantageous for some animals than for others. One possible answer comes to mind. The vegetarian monkeys that do not occur in mixed groups have comparatively small territories (*Callicebus moloch*, almost certainly the *Pithecia monacha* of the Caquetá and the Putumayo, and probably other forms) or they have large territories but move through them very slowly on the average (e.g., *Alouatta villosa*). Individuals of these species and populations seldom find themselves in situations with which they are not thoroughly familiar or have not had time to inspect carefully beforehand. Many of the more omnivorous or insectivorous types (at least *Saimiri* and *Cebus* spp.), by contrast, have relatively large territories and move through them very rapidly, probably because animal foods are relatively rare and often scattered. The Common Spider Monkey also has large territories and sometimes moves rapidly over long distances. It may do so because some of the large fruits it prefers are also dispersed (more dispersed than the leaves of howlers or the smaller fruits taken by smaller monkeys). Species of large territories and rapid movements may find themselves precipitated into unfamiliar situations rather frequently. It is not difficult to understand why they might have more need of extra scouts or sentinels than species of more sedentary, sluggish, or cautious habits.

More active groups probably tend to flush more small animals than do less active groups. This may explain why small hawks, which feed on insects and reptiles, are much more likely to associate with insectivorous or omnivorous monkeys than with the passive vegetarian types, and also why the active monkeys do not catch more of the flushed prey.

There is reason to believe that species and populations that tend to join others, or permit themselves to be joined, should be able to invade and maintain themselves in distant or "new" regions (major geographic regions, not limited areas such as the territories cited above) more frequently than others that do not participate in friendly heterospecific associations. Simply because they will often observe and benefit from the knowledge accumulated by the already established local residents of other forms, it should be comparatively easy for them to discover the locations of foods, predators, and whatever other features may be relevant to their survival in their new surroundings. Among some neotropical birds (Moynihan, 1962), mixed groups also are more common in marginal regions, especially those in which conditions are suboptimal, than in more central or favorable regions. Monkeys may illustrate these phenomena quite as well as birds. Central America, on the northern margin of the New World tropics, lacks several of the major types of monkeys that keep to themselves. So, apparently, does southeastern Peru, which is equally marginal in the opposite direction (Durham). The monkeys of the Guianas may be particularly apt to form mixed groups. These regions may also be less optimal for ceboids than is the upper Amazon. It would be a circular argument to claim that they are less favorable only because they are inhabited by a reduced number of species of monkeys, but they probably also have fewer species of other organisms, a narrower range of potential food sources.

Similar factors may be involved in another possible anomaly, the interspecific associations of the small sakis observed by Mittermeier in Brazil. He found these animals to be relatively rare, apparently less abundant than the *monacha* of the Caquetá or even the Putumayo. Perhaps the local habitats were unsuitable for them, marginal in an ecological if not geographical sense.

The various kinds of interspecific social behavior of New World monkeys—avoidance and other hostile reactions, as well as friendly joining and following—would be expected to be adjusted to different types and amounts of competition between species. On general logical grounds, one would assume that species that do not compete, or compete as little as possible for animals that occur in the same areas, would usually ignore one another. There appear to be examples among ceboids, e.g., *Saguinus fuscicollis* and *Cebus apella* in Amazonian Colombia. One would also assume that species that compete very strongly would tend to exclude one another from wide areas and entire regions. This may be the case with *Saguinus geoffroyi* and *Saimiri* in Central America. (Exclusion followed by accelerated divergence of habitat and food preferences probably accounts for the ecological separation of many other forms such as *Callicebus moloch* and *C. torquatus.*)

Presumably either of these extreme types of interspecific relationship can be transformed into the other in the course of time. It would be interesting to know the intermediate stages. Theoretical considerations and some data from studies of other animals (birds) would suggest that the following progression may be common as intensity of competition increases. When competition becomes slightly more than minimal, the species will tend to ignore one another in most circumstances, but will exhibit overt and active hostility toward one another occasionally. (If it is only desirable or necessary to drive off rivals infrequently, it may be worth taking the risk of fighting.) When competition is stronger, it may be advantageous for the competitors to join up with one another. (If you can't lick 'em . . .) When competition becomes stronger yet, it may become imperative to avoid one another. First by avoiding personal encounters while still ranging over the same areas at much the same times. Then by claiming exclusive territories or by elaborating some form of temporal segregation. (Separation by differential timing may have peculiar advantages, but it can work only when the species involved are not too numerous.) From the claiming of exclusive territories, there may be no more than a small step to complete allopatry.

It seems very probable that the process can also go in the opposite direction, through the same stages but in reverse order, and that the direction of change can be reversed repeatedly, with or without reaching the extreme conditions at either end.

Such as it is, the available information would seem to indicate that many New World primates conform to this rather hypothetical schema. But there are still further complications, special cases, and exceptions; some will be mentioned briefly.

Spider monkeys, at least *Ateles "geoffroyi,"* probably can afford to engage in interspecific fighting comparatively frequently, over objects or resources that might not stimulate overt aggression by other species, because they are very large and agile. They are more likely to win and less likely to be seriously injured during interspecific fights than are many other species.

Similar distortions may occur on a smaller scale. Individuals of *Saguinus geoffroyi* attack one another more often than do Squirrel Monkeys. They may, therefore, be more easily provoked to attack individuals of other species. If so, this would help to explain why they are avoided by squirrels, rather than joined and followed as are Squirrel Monkeys in some areas.

Squirrel Monkeys are more frequently and closely associated with a greater variety of commensals and companions of other species than are any other New World primates. Besides their reluctance to attack, this may be due to another feature of their behavior, or, rather, an adaptation that may explain their reluctance. (As noted above they are not really nonaggressive. They show many threat patterns among themselves.) The critical factor may be their manner of coping with potential predators.

Most species of ceboids that bother about predators have two different kinds of defensive reactions, both of which they use

quite frequently, opting for one or the other according to circumstances. They can try to avoid notice, and freeze or dodge
behind shelter when danger is found to be close. In other
words, they can try to be cryptic. The alternative is to discard
the possibility of hiding and to flee when danger is still distant.
A refinement is to perform mobbing in the hope of distracting
predators before escaping. Both series of patterns can be used
by animals of very different feeding habits and degrees of vulnerability. There are only a few species that always select one
method rather than the other. *Cebuella* and probably *Callicebus torquatus* try to avoid notice. *Saimiri* follow the opposite
course. In my experience, Squirrel Monkeys never freeze or
try to hide; consequently they usually can be found without
much difficulty. Observing them for any length of time is
another matter. They are extremely alert and excitable, and
often suspicious. Wherever they are hunted, they tend to retreat
at the slightest untoward sight or sound. Thus, one of the
principal disadvantages of acquiring extra companions, to enhance conspicuousness, may be minimal for them. The corresponding advantages may be relatively more important in consequence.

Squirrel Monkeys cannot have had a great deal of choice.
Their own troops are so large that they could hardly ever pass
unseen by a predator in their immediate vicinity. *Cebus apella*
may be a rather different case. Its own social groups are much
smaller than those of *Saimiri*. Other monkeys of similar or
larger size in some of the same areas, e.g., *Pithecia monacha*,
sometimes freeze and hide. So does *Cebus capucinus* in other
areas. But the *apella* of Amazonian Colombia resemble *Saimiri* in usually or always retreating directly and immediately.
This probably is an adaptation to the habit of associating with
Squirrel Monkeys, a precaution to ensure that the movements
of the two species continue to be coordinated. If so, then a
single mixed group of the region may include different species
in which inter- and intraspecific behavior have interacted in

very different ways to produce congruent results. This would be another example of the variety of relationships that can exist between the two classes of social responses, and of the "opportunism" that seems to have been characteristic of the evolution of both.

chapter five
COMMUNICATION SYSTEMS

The repertories of signals used by many New World primates
were discussed in Moynihan (1967). There have been further
studies of the signals of several forms, e.g. *Saimiri* and *Cebuel-la*, in more recent years. They have increased our knowledge
by many factual details, without, I think, altering our impres-
sion of the general situation. Thus, some of the following notes
will be brief and recapitulatory. Others will be concerned with
certain theoretical and evolutionary questions.

The various signals of ceboids, like those of most other mammals, can be grouped into four categories: tactile, olfactory,
visual, and acoustic. These may intergrade or overlap. A single
pattern can be perceived by several senses simultaneously. It
seems likely, however, that one sensory modality usually pre-
dominates at any given time (different modalities in different
cases).

All behavior patterns that convey information from one individual to another should be considered signals, even though
some may subserve many other functions as well. Certain patterns seem to have become specialized in form or frequency
expressly to convey information. They may be said to be "ritualized" and may be called "displays."

All patterns cited below are typically adult (possibly also
performed by younger individuals when they behave like
adults) unless specifically stated otherwise.

Tactile Signals

Single species or small groups of species may have one or
more distinctive signals that are mediated by the sense of touch
and are not shared with other monkeys. Most of these are relatively minor. There is only one tactile display that is widespread among ceboids: Allogrooming, the grooming of one in-

dividual by another. It is known to be performed by all the species studied (the first record for *Cebuella* is in Christen, 1974).

Allogrooming probably has been derived from self-grooming in the course of evolution. Grooming as a whole, both types included, can involve the use of the hands, feet, or teeth. The hands may be used to scratch and to curry or comb the fur or hair. The feet are used to scratch. Both hands and feet may be used to remove ectoparasites, as can the teeth, which also can be used to pick off bits of dead skin. In all species, scratching with the feet seems to be essentially confined to self-grooming. There is considerable variance in some other aspects of form. Such species as *Saguinus geoffroyi, Aotus,* and *Callicebus moloch* tend to use both hands and teeth during Allogrooming, but only the hands and feet without the teeth during self-grooming. *Pithecia rubicunda* tends to use both the hands and the teeth with the feet during self-grooming, but only the hands during Allogrooming.

The Allogrooming patterns of different species also differ in such features as frequency of performance, motivation, and results obtained.

The primary social function of Allogrooming in all species seems to be to break down barriers. All adult or even half-grown monkeys show some reluctance to allow themselves to be touched. Presumably this is insurance against sudden disputes or attacks, yet it has its drawbacks. Certain social reactions, most notably copulations, can hardly be performed without personal contact. It may, therefore, be advantageous to have some pattern(s) that can facilitate physical contacts whenever desirable without necessarily dissolving all distrust in all circumstances. This is what Allogrooming seems to be able to do for many species.

It seems to be effective because it can provide great satisfaction to the individuals being groomed. A potential "groomee" may show signs of nervousness or alarm when first approached by a "groomer" (a point of some importance—see below). But if the Allogrooming can be started, then there may be an immediate relaxation and lowering of tension. The groomee may go quite limp and assume an "ecstatic" expression with the eyes partly or completely closed. It may even turn or shift to present different parts of the body to be groomed in succession. No observer could doubt that it is happy. One never sees a monkey looking happier.

Of course, Allogrooming may also subserve a nonsocial function. It should help to keep the skin and hair in good condition, much like self-grooming, but this may be a secondary by-product. There does not seem to be any *significant* correlation, among either individuals or species, between frequency of Allogrooming or being Allogroomed and susceptibility to diseases or deterioration of skin or hair, although possibly some animals that are not groomed by others may have to groom themselves more often in compensation (again see below).

One extreme is represented by *Aotus*. In this species, Allogrooming is closely associated with copulation. Among adults, it seems to be restricted to situations in which copulations can be expected to occur. Thus, it would appear to be an expression of sexual motivation, probably a special type of such motivation, something that could be called a copulatory drive.

Callicebus moloch represents the other extreme. Its Allogrooming is a general social or gregarious pattern, performed by almost all individuals of the same social group, not only actual or potential mates, at all times of the year. This also seems to be true of the Allogrooming of *Cebus capucinus* (Oppenheimer, op. cit.), *Pithecia monacha*, and *P. rubicunda*.

The corresponding behavior of *Saguinus geoffroyi* is performed much less frequently, but is not really rare in absolute terms. It is complex. Adults of the species may perform general social Allogrooming occasionally. Such performances are outnumbered by Allogrooming between mates. These, in turn, are most common during the breeding season. They are not, however, confined to copulations or copulatory situations; they must be produced by sexual motivation, but many of them are less likely to be expressions of a strictly copulatory drive than of some sort of pairing tendency.

As an adult pattern, the sexual version of Allogrooming probably is more primitive than the general social version. All or most of the Allogrooming of adult mammals of other orders is sexual (Andrew, 1964).

It may be wondered why the primitive sexual pattern has been extended to other social situations in some species of ceboids, or why it has not been extended further in such forms as *Aotus* and *Saguinus geoffroyi*. The answer may lie in the availability of possible substitutes. Allogrooming must (or would) provide advantages for many, perhaps all, monkeys. But the obvious nervousness of some individuals when approached to be groomed would seem to indicate that the whole business can be rather tricky. The pattern may be soothing and reassuring as soon as it has been begun, but the beginning itself may be difficult, carrying some appreciable risk of provoking panic with subsequent escape or fighting. Thus, one might expect Allogrooming to be used by species only when and insofar as they do not have other social patterns that might procure the same benefits or similar advantages with less uncertainty.

Aotus and *Saguinus geoffroyi* differ from *Callicebus moloch, Pithecia* spp., and all the larger ceboids in usually or always sleeping in holes in trees under natural conditions. All individuals of the same family or group of tamarins or Night Monkeys usually sleep in the same hole. More often than not, they must be touching one another while they sleep. (Even this form of contact can be preceded by many hesitations. At least in captivity, individuals are almost always reluctant to enter the sleeping boxes provided for them. They may go in and out half a dozen times. They may also fight inside the boxes, but they eventually settle down as best they can.) Monkeys that do not use holes tend to remain slightly dispersed when sleeping. One can understand, therefore, why they might need to extend or elaborate other patterns to encourage contacts, and why individuals of species that are more often together during periods of repose would not need to do so. Night Monkeys and tamarins probably can afford to confine their Allogrooming to circumstances in which the closest possible approach is peculiarly important, simply because they usually have other contacts sufficient to facilitate or permit most of their other social reactions.

There is a general tendency, among most adult and subadult ceboids, for subordinate individuals to groom dominant ones more frequently than the reverse. But the rule does not apply to *Pithecia monacha* or *P. rubicunda*. Among older individuals of these species, it is more common for dominant individuals to groom subordinates. Possibly this is an extension of parental behavior. Parents groom their infants in many or all species.

Alouatta villosa and *Saimiri* perform Allogrooming even less frequently than *Saguinus geoffroyi*. Their Allogrooming is both relatively and actually rare. This may be because they are so gregarious and cohesive among themselves. Individuals tend to remain so near to others at all times that further patterns to stimulate closer relations might be supererogatory. Both species also have highly specialized and distinctive precopulatory displays, i.e., Tongue-pumping in *Alouatta villosa* (Carpenter, 1934) and Penile Erection in *Saimiri* (see above and Ploog and Maclean, op. cit.).

As already noted in Chapter 3, Panamanian howlers suffer greatly from skin parasites. Their vulnerability may be due more to the poor development of their self-grooming than to the deficiency of Allogrooming. Squirrel Monkeys perform a fair amount of self-grooming and do not seem to have more skin parasites or diseases than most other monkeys.

According to Neville (1972b), *Alouatta seniculus* does more Allogrooming than *villosa* and, significantly enough, less elaborate Tongue-pumping.

It is evident that adult Allogrooming has evolved in opposite directions in different stocks of ceboids. Starting out as a sexual display, it has been generalized in some cases and suppressed and replaced in others. Both changes probably occurred independently in several different lineages.

Olfactory Signals

The behavior of New World monkeys would suggest that their sense of smell is acute, much more so than our own. They appear to rely upon it to a very considerable extent when inspecting previously unfamiliar objects or assessing the possible value of different kinds of foods, and probably also for recognition of individuals.

Unfortunately, this is a subject that has been very little studied. To my knowledge, no one has tried to analyze the physiological or neurological parameters, or identify the odors perceived and reacted to.

There are many indications, nevertheless, that the majority of species have patterns that should be called olfactory displays. Some are conspicuous to the eye as well as the nose. Of these, the most widespread are various kinds of Rubbing. Special secretory glands on one or more parts of the body, such as the chest or the genito-anal region, are rubbed against the substrate, usually branches or twigs in the wild, in more or less stereotyped ways. The distribution of glands in many species has been described by Epple and Lorenz (1967). My own observations would suggest that the majority of Rubbing performances are partly or purely hostile. Many of them seem to be concerned with claiming and marking territories.

Visual Signals

These are much less problematical than the olfactory variety. They include many displays and a whole host of unritualized patterns. The displays can be divided into three groups by their morphology: (1) pilo-erection, hair-raising, displays with some contradictory or counteracting forms of smoothing; (2) facial expressions; and (3) gross movements and postures.

Pilo-erection patterns occur in many species. They are most elaborate and diverse in the tamarins and marmosets, which have elongated or otherwise specialized tufts or ruffs of hair.

All species, with the possible exception of *Aotus*, have several distinctive facial expressions, but the variety is not often very great. New World primates tend to be rather poker-faced on the whole. Among the species that have been studied in detail, it is only the capuchins, Common Spider Monkeys, and the Woolly Monkey that have been found to have many expressions and to ring the changes upon them frequently. (Preliminary observations would suggest that *Pithecia rubicunda* may have as much potential flexibility. In captivity, however, it tends to limit itself to a comparatively restricted range.)

There would appear to be a crude negative correlation between pilo-erection displays and facial expressions. The former are most complex in small species. The latter are most complex in some, not all, large species. This may be explained by differences in visibility. Tamarins and marmosets are so small that their expressions are difficult to decipher at a distance. The raising of elongated tufts and ruffs of hair may increase visibility. But such specialized structures may be difficult or onerous to maintain in good condition. Rather like Allogrooming, they may be selected for only when no alternatives are available.

The gross postures and movements are heterogeneous, exceedingly diverse in form. They can involve the whole head and body, or the head alone, or large parts of the body, or the limbs, or the tail.

Examples of all three categories are illustrated in Figures 1 through 27. The series is biased in favor of hostile facial expressions, much the most interesting of the patterns to draw. The accompanying legends give details.

Each category includes patterns of very different significance, designed to transmit very different messages, hostile, friendly, sexual, etc. Highly aggressive and extreme alarm patterns have been the most conservative during evolution. They tend to occur at critical moments and changes in them may be particularly dangerous.

There are three kinds of ceboids that seem to place less reliance upon visual displays than do the other forms. These are *Aotus, Saimiri,* and some or all of the howlers (certainly *Alouatta villosa* and probably other species of the genus as well).

The Night Monkey has no pilo-erection patterns, probably no special facial expressions (unless the opening of the mouth during vocalizations can be considered such), and relatively few ritualized postures or gross movements of the head or other parts of the body. Nor does it seem to use unritualized movements as signals to any great extent. This is obviously correlated with nocturnality; visual patterns are difficult to see at night.

Alouatta villosa and *Saimiri* also lack pilo-erection displays, have only a few facial expressions, and do not show more than a moderate variety of other ritualized movements and postures. But they probably use unritualized visual signals very frequently indeed. They seem to pay close attention to all the unritualized movements and intention movements of their companions, and to derive some of the same social information from them that other species may derive from ritualized displays. Many unritualized patterns are less conspicuous than most displays. It seems likely, however, that howlers and Squirrel Monkeys can use inconspicuous signals with more assurance, and greater success, than most other species because their groups or sections of groups are so cohesive that members often are near enough to one another to detect even the slightest cues.

If so, then the peculiarities of the visual communication systems of *Saimiri* and *Alouatta* are direct consequences of their distinctive social organizations in the same ways as are the peculiarities of their tactile systems. (There must have been convergence during the evolution of these arrangements. *Saimiri* and *Alouatta* are quite different in most characters apart from some general aspects of social behavior. Several of the displays they do perform are not homologous, viz., the pre-copulatory patterns cited above.)

The situation can be summarized in a few sentences. *Aotus, Saimiri,* and *Alouatta* all have comparatively simple (impoverished) repertories of visual displays, but not for the same reasons. A great diversity of such displays would be useless to *Aotus*; it is unnecessary for the other two genera. See also below.

Acoustic Signals

Most of the social sounds produced by New World primates are vocalizations.

There is a "typical" ceboid vocal repertory. With variations, it is employed by adult *Aotus, Pithecia monacha, P. rubicunda, Callicebus moloch, C. torquatus, Lagothrix,* the Common Spider Monkey, *Saimiri,* and the four capuchins. Presumably adults of the remaining species of *Pithecia, Callicebus,* and *Ateles* also use it.

It includes several different kinds of high pitched sounds, most of which can be called Squeaks, Whistles, or Trills. They usually seem to be hostile when uttered by adults of most of the species that exhibit the full range. (*Aotus* is a partial exception. The Squeaks of adults of this species are sexual.) All or most of the hostile patterns in this category are low intensity, produced by relatively weak motivation, and are not very aggressive. Brief single notes, such as Squeaks, tend to be lower intensity than longer vocalizations. Rapid series of brief notes, such as Trills, often are produced when escape motivation is relatively stronger than in more prolonged or continuous notes, such as Whistles.

Closely related to these patterns are Screams. They also tend to be high pitched. They are hostile but very high intensity indeed, and are used only in extreme emergencies.

A typical repertory always includes low pitched notes in addition to high ones. Many low notes can be described as Barks or Grunts. These are as purely hostile as most of the adult high pitched sounds, but seem to be more aggressive on the average. Probably none of them is quite as low-intensity as many Squeaks. In some species, the most exaggerated and highest intensity patterns of this low pitched group are prolonged into Roars (the equivalent of the "howls" of howlers).

Barks, Grunts, and Roars sound rather harsh. Other low pitched patterns are clearer or smoother, more melodious, plaintive, or whining in quality. They can be called Hoots, Moans, and Wails. All or most of them are produced by thwarting of some nonhostile, sexual or friendly, motivation, or by conflict between such motivation and hostile tendencies (perhaps no more than a type of frustration of the friendly or sexual drives).

A few species have some vocal patterns of medium pitch. This category is heterogeneous. One medium pattern of one species may be homologous with notes that are low pitched in most other ceboids, while another medium pattern of the same or another species may be homologous with notes that are usually high pitched. A variety of patterns from typical repertories are illustrated in Figures 28 through 38.

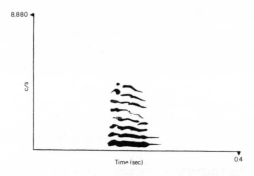

Figure 28. A drawing of a spectrogram of a Gruff Grunt by an adult Aotus. *The vertical axis indicates cycles per second, the horizontal axis, time elapsed. Compare with Figures 29, 30, and 32. All illustrations of the sounds of Night Monkeys are after Moynihan (1964).*

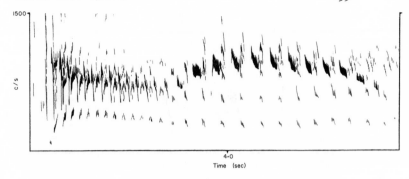

Figure 29. *The first part of a long, full "Song" by an adult* Callicebus moloch. *The initial note may be a typical "Chuck" (see also Figure 31). The following sounds are Resonating Notes (first abbreviated and then prolonged). These probably are related to the Barks and Grunts of other species. All illustrations of* moloch *sounds are from Moynihan (1966).*

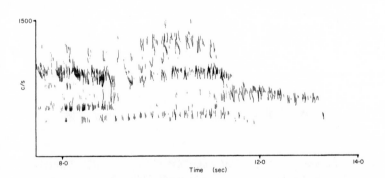

Figure 30. *The last part of the same "Song." This follows directly upon the phases shown in the preceding figure. It consists of Pumping Notes, again related to the Barks and Grunts of other species.*

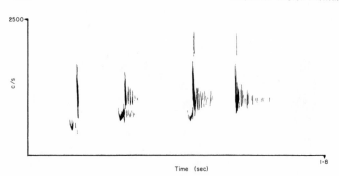

Figure 31. Muffled "Chuck" Notes by an adult female Callicebus moloch. *These seem to be medium pitch to human ears, but they may be related to higher or lower alarm notes of other species. Compare with Figures 42 and 44.*

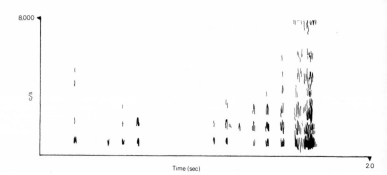

Figure 32. A series of notes uttered by an adult Pithecia monacha. *Most of the sounds are Bubbling Notes, related to the Grunts or, perhaps more probably, the Moans of other species. The last sound is a more definite Grunt. See also Figures 34, 35, and 40. The distances between the notes in this and all the other illustrations indicate the real intervals.*

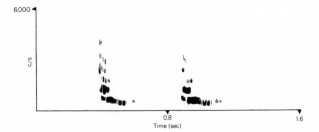

Figure 33. Two Hoots by an immature male Aotus. *These are very reminiscent of some sounds uttered by owls, surely not coincidentally.*

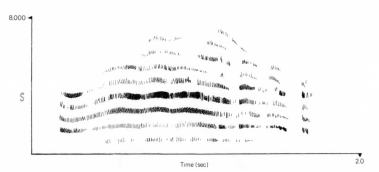

Figure 34. A Wail by an adult Pithecia monacha. *This is one of the most extreme of Moanlike sounds. It also exemplifies a feature that recurs in many different vocalizations of many species. It sounds like a single note to human ears, but spectrograms reveal that it is composed of a series of separate pulses.*

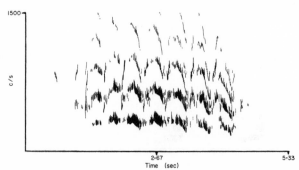

Figure 35. A series of Moans by an adult Callicebus moloch.

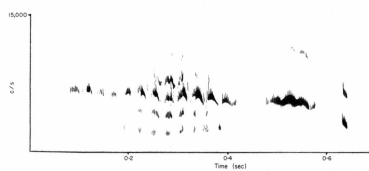

Figure 36. High pitched vocalizations by a juvenile Callicebus moloch: *a Trill followed by a brief Whistle followed by a single Squeak.*

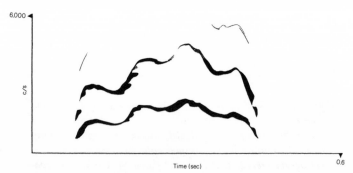

Figure 37. A brief Scream uttered by an adult Aotus. *Compare with the next figure.*

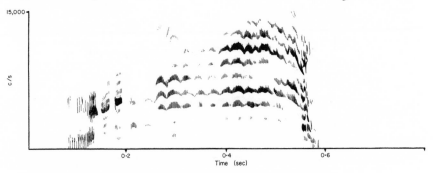

Figure 38. A brief Scream by a juvenile Callicebus moloch.
*The initial pulses of sound may be transitional between
Grunts and the true Scream. The terminal pulses are almost
"Chucklike."*

(Terms such as "high pitched" and "low pitched" are rather
crude, only approximations. Patterns have been ascribed to
particular classes, in this account, on the basis of their sound
to human ears. The illustrations reveal that the actual ranges
of frequencies, cycles per second, can be similar or broadly
overlapping in notes that have been put in different classes.
Such notes differ in pitch because of differences in the loudness
of some of the frequencies within them.)

Two or three kinds of ceboids have vocal repertoires that do
not conform to the typical outline. These are the tamarins and
marmosets on the one hand and (again) *Alouatta* on the other.

All or almost all the vocalizations of tamarins and marmosets
are high pitched. This feature may be less significant than
might be supposed on first hearing. Tamarins and marmosets
do not seem to lack homologues of patterns that are low
pitched in other species; they have merely transposed them into
a higher key. Some patterns of two species of tamarins are illustrated in Figures 39 through 47.

The howlers (certainly *Alouatta villosa* and apparently *A.
seniculus*) have gone in the opposite direction. All their vocalizations are more or less low pitched. They may have lost the
homologues of patterns that are high pitched in other species.

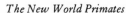
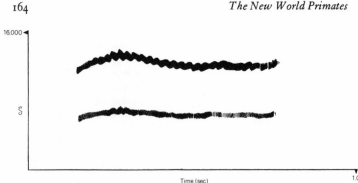

Figure 39. A Whistle by an adult Callimico. *This verges upon a Trill. Compare with Figures 36, 37, and 38. I am indebted to Dr. J. K. Hampton for permission to record captive individuals in his laboratory.*

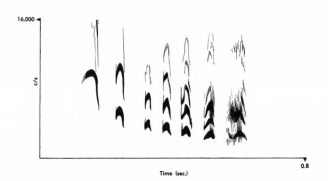

Figure 40. A Trill by an adult Saguinus geoffroyi. *The last two bursts of sound are partly intermediate between typical Trill Notes and Rasps. See also Figures 35, 36, 41, 42, and 43. All illustrations of the sounds of this species are from, or after, Moynihan (1970a).*

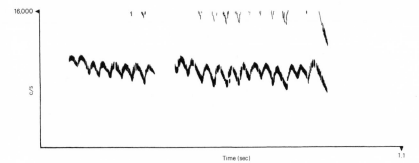

Figure 41. Two soft Trills by an adult Callimico.

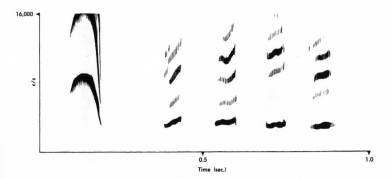

Figure 42. A series of sounds uttered by an adult Saguinus geoffroyi. *A Sharp Note (an alarm note) followed by a brief Twitter (composed of Short Whines, modified Whistles possibly related to the Moans of other species).*

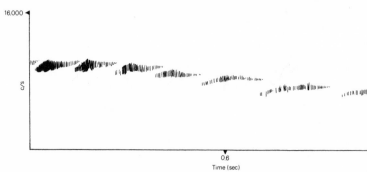

Figure 43. The end of a very long, faint Trill by an adult
Callimico. *This species has a variety of Trill-like patterns.*
Obviously the one shown here is very different from the ones
illustrated in Figure 41.

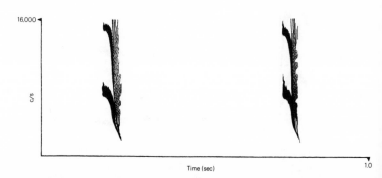

Figure 44. The lower parts of two Sneezing Sharp Notes by
an adult Saguinus geoffroyi. *These are combinations of ordi-*
nary Sharp Notes (see Figure 42) and typical sneezes.
Sneezes are used as signals by several ceboids.

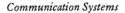
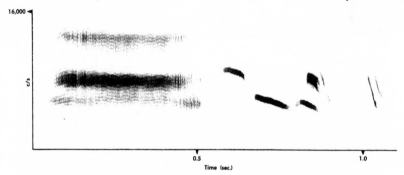

Figure 45. *Vocalizations by a young* Saguinus geoffroyi. *An Infantile Rasp followed by one, or one and a half, Infantile Squeaks.*

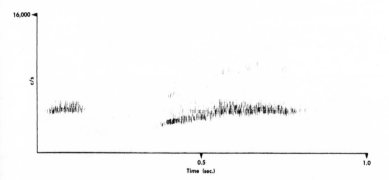

Figure 46. *Two long Rasps by an adult* Saguinus geoffroyi. *All or most of the adult Rasp patterns are related to the screams of other ceboids, but some of them are among the lowest pitched of the vocalizations of the species.*

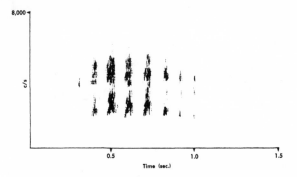

Figure 47. A Broken Rasp by an adult Saguinus geoffroyi.
Compare with Figures 45 and 46 (also 32 and 40).

If so, the loss of some patterns has been accompanied or followed by a modest proliferation, division into more types, of
the survivors.

The adaptive values of different pitches are fairly obvious.
High pitched sounds do not carry as far as low pitched ones
(especially in environments such as forest and scrub where
there are many obstructions to scatter or deflect sound waves).
In the cases of species with typical vocal repertories, it may be
just those patterns that are usually employed for short distance
communication, or that might be too dangerous to broadcast
widely, that tend to be high pitched, while many of the low
pitched notes certainly are used for long-distance communication. The small tamarins and marmosets may be potentially
vulnerable to a particularly wide range of predators. The fact
that their notes are high pitched may help them to avoid attracting the attention of predators whenever possible and desirable, i.e., whenever mobbing is not necessary or advantageous.

As usual, there are apparent discrepancies that need to be
explained (or explained away). Two may have been noticed
already. The vocalizations of howlers are low pitched and
should be suitable for long distance communication, but it has
been suggested that the characteristic features of their tactile

and visual signal systems are adaptations to short distance communication, all that is really needed in their closely knit groups. The last phrase is the key. There is communication between as well as within groups. Among howlers, the most important intergroup signals are the spectacular Roars. The specialization of the hyoid apparatus, which is diagnostic of *Alouatta*, would appear to be designed to augment the strength and/or length of Roars. Once evolved, however, it may have impeded or prevented the production of high pitched notes. There may be a physical limitation: high pitched sounds may not be able to get through the resonating chambers of the hyoid in usable or recognizable form. Thus, the howlers may have been condemned to utter low pitched sounds during intra-group reactions even though they are not ideally adapted to the circumstances. In the case of *A. villosa* at least, the animals compensate by uttering many of the low pitched sounds within groups very softly. Another factor that could be relevant in this connection is that howlers may be better able to afford consistent lowness of pitch than many of their relatives because they are larger and, therefore, presumably exposed to fewer predators. *Callicebus moloch* individuals continue to utter some high pitched sounds despite their apparent invulnerability. They have less specialized hyoids than do howlers. Selection in favor of retaining high pitched sounds may have been particularly strong in their case because of their peculiar type of intraspecific crowding. Pairs and family groups of *moloch* often are much closer to one another than are troops of *Alouatta* or the basic units of any other species. Their use of short range patterns may help to avoid mixing or confusion among territorial neighbors.

Other differences between species involve frequencies of performance. Some species are much more consistently noisy, i.e., they utter notes much more readily, than others. This may also be correlated with predation. Most small species are less frequently vocal on the average than most large species. Most highly gregarious species, which are inevitably conspicuous enough as it is, are less vocal than less gregarious species of similar size. The tiny *Cebuella* is the least noisy of all the species studied in the field. *Saimiri* is the smallest of the highly gregarious ceboids and also tends to vocalize comparatively rarely. (The fact that individuals of this species cannot usually pass unseen by nearby predators—see Chapter 4—does not mean that it could not be advantageous for them to remain unheard when out of sight.) The well-protected *Callicebus moloch* may be the most frequently vocal species possessing a typical repertory. It also is one of the largest diurnal species that is not highly gregarious. (*Pithecia monacha* is less continuously noisy although it is even larger and almost equally nongregarious. This may be because it does not stick so tightly to the densest vegetation.) Among the species with nontypical repertories, it is the howlers that are the most repetitive as well as loudest of vocalizers. (This indicates that the quietness of Squirrel Monkeys is unlikely to be due to their reliance upon unritualized visual signals.)

In *Aotus*, the major vocal patterns of adults are quite discrete. Night Monkeys do not usually utter many intermediate patterns. In *Callicebus moloch*, by contrast, intermediate notes are both relatively and actually very common. Many of the adult patterns of the species seem to intergrade through a perfect continuum. Epple (1968) suggests that *Callimico* may be similar to *Callicebus moloch* in this respect. Some ceboids are less extreme. The organization of the adult vocal repertory of *Saguinus geoffroyi*, for example, is almost exactly half way between that of *moloch* and that of *Aotus*. Although the details are quite different, the corresponding repertory of *Cebus capucinus* also seems to be partly intermediate (Oppenheimer, 1973). Probably other species will be found to show other variations.

Some of the differences may be explained as the products of, or compromises between, objectives that are not entirely compatible. Signals should be as clear as possible; they should also contain as much information as possible. Unfortunately, the two objectives are difficult to reconcile in the same message at the same time. Discrete vocal patterns are clear and perhaps emphatic, but they are crude. They cannot convey many fine shades of meaning. Intergrading patterns have the reverse qualities. They can express ambivalence and subtle distinctions, but probably only at the risk of some ambiguity.

The nocturnal *Aotus*, sometimes exclusively dependent upon sound for communication, has to ensure that its signals cannot be misinterpreted. Diurnal forms may have more leeway. They can often afford to transmit difficult and complex messages because their listeners receive visual information, from the whole surrounding situation, to help with the decipherment. Of course, visual information varies with the nature of the environment. Marler (1973 and elsewhere) has suggested that, among diurnal Old World monkeys and apes, there are direct correlations between intergrading calls and open habitats, and between discrete calls and closed or cluttered habitats. The available, incomplete, evidence would seem to indicate that similar correlations occur among ceboids.

Another factor may also be relevant. Different species *use* their vocalizations differently. Again this is most obvious in the cases of *Aotus* and *Callicebus moloch*. Many vocalizations of adult Night Monkeys seem to be very precise signals. They induce the same or very similar responses by all receivers of the same social class (age, sex, physiological condition) in most circumstances. The responses usually are fairly rapid. The vocalizations of adult *moloch* often are less definite as releasers of "commands." It is quite common for the same signal to provoke different responses in different circumstances, and many responses may be very long delayed. Perhaps the vocalizations of *moloch* "merely" provide information that a receiver can choose to ignore, file away for future reference, or react to in any one of several ways, depending upon other information obtained or available from other sources.

The normal ontogeny of vocalizations is difficult to study. Soft calls of infants may be inaudible to human observers in the field. Data derived from laboratory animals may be abnormal or unreliable (see also below). Pola and Snowdon (in press) have recently described some of the complexities that may occur in the development of sounds among Pygmy Marmosets in captivity. It is at least evident, nevertheless, that infant and juvenile ceboids do not have exactly the same vocal repertories as their parents. The young of some species have uniquely infantile calls and notes, all or most of which are distress signals. The young of other species do not have such distinctive patterns, or use them only rarely. They utter some adult-type calls but not others, i.e., their repertories are narrower than those of older individuals.

It would seem that special infantile distress calls are absent, or least developed, in the most gregarious and mobile forms such as *Saimiri* and the Common Spider Monkey. Perhaps the young of these species meet too many adults that are not prepared to be indulgent, or encounter them too frequently, to make it safe to sound like anything but other normal (if not very fluent) adults and responsible members of the community. Another example of the dictum that children should be seen and not heard.

Summary and Implications for the Evolution of Languages

With all their differences, the signal repertories of ceboids have remained conservative in basic form. Or, rather, their components have. The majority of the patterns of any given species are homologous with patterns of some or all of the other species. The correspondence is not always one-to-one, but the relationships are clear.

Ritualized signals also have been conservative in another respect. All the species of ceboids that have been studied at length have been found to have roughly similar total numbers of qualitatively distinct kinds of "major" displays (excluding intermediates and minor variants). The numbers seem to range from something like 16 to approximately 35. This range might appear to be substantial, but it is much less than would be physically possible, and the extremes are well within the same order of magnitude. Almost every species fails to perform ritualized homologues of some displays of its relatives even though it has the physical equipment to do so. Species that are more intelligent, more gregarious, or both, may tend to have a few more major displays than do less gregarious, less intelligent ones, but the difference is slight.

Limitations of the numbers of kinds of displays are not peculiar to New World primates. They also apply to the repertories of many other animals, including birds and fishes. See Moynihan (1970b).

It may be interesting to compare the communication systems of monkeys and apes with those of man. The subject is admittedly difficult, but something may be gained by a discussion in the simplest possible terms, even at the risk of oversimplification. (For purposes of comparison, it usually will be convenient to treat all the monkeys and apes of different regions and continents as a unit. The known data suggest that all or most of them have repertories of similar size to subserve similar functions in similar ways. Only a few species will deserve individual mention to illustrate particular points or exceptions.)

Altmann (1967) has argued, very plausibly, that most of the general design features or basic principles of human communication systems, as identified by Hockett (1959 and 1960), e.g., choice of channels, address, rapid fading, interchangeability, semanticity, arbitrary denotation, etc., are equally typical of the corresponding systems of many other primates. This does not exhaust the list of resemblances.

Most of the nonverbal signals of man, acoustic patterns such as laughing and sobbing as well as many facial expressions and movements, are no more distinctive than those of most other terrestrial vertebrates or fish or insects or cephalopods. Some of them are species-specific, but they are not unconventional.

The real problems accumulate with verbal messages. They

appear at many levels of both fact and theory. There has been some dispute as to the propriety of employing the term "language" for various types of animal communication. From the point of view of pure and formal logic, it is perfectly defensible to restrict the term by definition to human verbal systems. This does not seem to me to be a very useful definition. It is at least conceivable (and perhaps proven in at least one case—see below) that other organisms on this or some other planets, or some machines, could develop or be programmed to produce some distinctive system that it would be desirable to call language. Even the most "restrictionist" of linguists constantly refer to *human* languages as if there could be some other kinds. One of the best known definitions of language by a restrictionist is "a set (finite or infinite) of sentences, each finite in length and constructed out of a finite set of elements" (Chomsky, 1957). If I interpret this correctly, and if the term "sentence" can be applied to sequences of signals perceived by any or all senses, which seems only reasonable, then the definition is almost all-inclusive. It describes the communication systems of man and also those of all other primates and many other animals down to the level of Paramecia, if not lower, and even those of some plants (e.g., the cues provided for insect and other animal pollinators).

Thus, it would appear to be justifiable to extend the term "language" to include all complex systems of communication. All are "propositonal, syntactic, and purposive" (Chomsky, 1967). Certainly all monkeys and apes have a command of something like a grammar. The meanings of their signals are altered by modifications of form and changes of sequence, both of which follow regular rules and can be predicted. Also the morphological components or building blocks of the repertories of most other mammals and birds are as varied and elaborate as the corresponding elements, the phonemes, of any human verbal system.

Granted all this, there is still no doubt that human verbal languages must be considered distinctive. They are distinguished by extreme developments and combinations that are unusual—not necessarily unique—among the various communication systems known at the present time. Some of the human peculiarities are so obvious that they have impressed zoologists, e.g. Simpson (1969), as well as linguists. Most of them have been discussed at enormous length. They all have histories and consequences. It seems to me, however, that only a few basic adaptations, perhaps no more than one, can have been crucial to the *origin* of the human type of verbalization. This might be worth emphasizing.

Modern human verbal systems can be used to transmit more information, in more complex codes or ciphers, than can the signals of any other organisms. Modern men also tend to be more intelligent than other organisms. Both abilities or attributes are late products of evolution. They may be causally related to one another. They do not, as such, throw much light upon what might be called the process of humanization. Human methods of communication presumably are derived from more typical animal systems. They must have been simpler at some earlier stage, when verbalization first began to be selected for. Moreover, within the human species, variations in intelligence are far from being perfectly correlated with variations in the capacity for speech. Even some "idiots" can speak in a human fashion (Lenneberg, 1964 and 1967). Relationships between intelligence and the development of language are not always close or direct. The ceboids demonstrate this as well as man. There are no qualitative differences between the communication systems of some clever forms, such as the Common Spider Monkey, and some stupid ones, such as *Alouatta villosa* and *Saguinus geoffroyi*.

Leaving aside content (for the moment), the most conspicuous distinction of human speech, perhaps the only one that is almost a matter of kind rather than degree, is the method of acquisition during ontogeny, the individual life cycle. Verbal languages contain more *learned* components, at some level, than do the communication systems of other animals. Doubtless the capacity to learn to verbalize is partly inborn. There seems to be a propensity to recognize and react preferentially to some "deep structure," a particular grammatical schema. The learned elements may have to conform to certain general rules of organization in order to be effective. There is still a great deal of flexibility in other respects. Each individual human child has to be taught much of the speech of its own society. Whatever its predispositions, it still has to learn how to arrange particular sounds or phonemes to make words and sentences. The phonemes would seem to be nearly meaningless by themselves (not entirely—harsh sounds probably tend to induce different responses from smooth ones). Different populations of human beings have developed different verbal languages by rearranging many of the same basic components to form very different looking or sounding entities. The resulting products are remarkably diverse at an immediate level, as different as English and Chinese and the Click languages of African Bushmen.

Individuals of many other species have to learn the proper orientation and timing of their signals, often by trial and error, but both the majority of the structures and the variations thereof are much more largely "innate." They are species-typical in all important respects, and tend to develop in the same or similar ways in all normal circumstances, the full ranges of physical and biological environments that permit survival of the individual and the species. It is true that some birds have developed local dialects (see, for instance, Marler and Hamilton, 1966, and Nottebohm, 1970). Some species of monkeys, e.g., *Aotus*, also appear to have dialects. But the differences between such dialects are very much smaller and involve fewer patterns, a much smaller percentage of the total repertory, than the differences among human verbal systems. They are hardly commensurable.

Barring physical defects, any human being who can learn one verbal language is potentially capable of learning to use any other. The factor of use is significant. Many animals of very diverse groups can learn to interpret the signals of other species. There are many well known examples. Birds in mixed flocks may come to react appropriately to the notes and calls of companions of other species. Anyone who has kept or watched a pet knows that it can understand some of its master's signals. But animals that comprehend the signals of other species almost never imitate them intelligently or perform the same patterns in return to exert a reciprocal influence. There may be a few exceptions among birds, some parrots and corvids (K. Lorenz, 1952), but they are rare and restricted to unusual situations.

We have some idea of what can happen, or fail to happen, when ceboids are given opportunities to learn "new" language elements. Some individuals of *Saguinus geoffroyi* develop remarkably aberrant behavior patterns when subjected to unusually strong social stresses in the abnormal conditions of captivity. These patterns have been called "quirks" (Moynihan, 1970a). They are special movements and postures, often quite elaborate in form. Comparable patterns can be induced in *Saimiri* and *Ateles* "*fusciceps*," and are even more common and diverse among captive *Cebus capucinus* (M. Bernstein, personal communication). The quirks of all species are as conspicuous as visual displays, but they take different shapes and are not species-typical (or not absolutely so). Different individuals of the same species can develop very different quirks. Quirks have not been seen in the wild under more natural conditions (Oppenheimer, 1968, and personal investigation). Presumably wild individuals are seldom frustrated enough to go really queer. When and where quirks do occur, they are potentially capable of serving as signals. They indicate the state of the performing animal, and they may reveal some of the factors responsible for this state. They can convey appreciable information to human observers. But they usually have very few effects upon other monkeys. According to M. Bernstein, the quirks of *capucinus* individuals do not usually provoke overt, positive, responses by other individuals of the same species or other species of the same genus. Most of them are not contagious, not passed on by imitation or tradition.

As capuchins can react rapidly to other kinds of unfamiliar stimuli, new animals and objects that might be edible or useful or dangerous, and often do so appropriately, their apparent lack of responsiveness to quirks would seem to be a definite refusal to notice, the result of some mental block. They may have been subjected to strong selection pressure *against* learning new social signals. At least against learning them very quickly. The quirks of the species seem to be outlets for frustration, doubtless satisfying to the individuals that perform them and thereby soothing, helping to reduce tensions, but that is all. The learning of new signals may have been selected against during the history of *capucinus* and other ceboids for several reasons. One is obvious: "innate" signals are usually adequate in the normal biological and physical circumstances. Reliance upon acquired or individual peculiarities would be an unnecessary refinement as long as similar circumstances continue to be normal. Unnecessary characters are always disadvantageous insofar as they use up resources that could be employed for other purposes.

Then why should humans differ from *Cebus* and other primates in always learning such large parts of their languages?

Primitive men and their ancestors probably did not have to convey much more *general* information to one another than do many other primates. They do not seem to have inhabited more complex environments. The human stock originated in the Old World tropics in forested or mixed habitats (see Tattersall, 1969a and 1969b), where other primates have lived for many millions of years. There is no reason to suppose that protohominids were more numerous or highly or tightly gregarious than many other primates of the same regions that have not evolved verbal languages. Fossil remains of early men and their ancestors are comparatively rare. Populations of existing hunting and food-gathering human groups, which may have retained or re-adopted primitive or pseudoprimitive habits, are always small and usually dispersed. G. McBride (personal communication) and others have suggested, alternatively, that the original human groups were unusually variable. They can hardly, however, have fluctuated more than do some populations of *Saimiri* and *Ateles*. Protection factors may be discounted for similar reasons. The ancestors of man probably were not threatened by more dangers than some of their relatives. It may be debated when they became terrestrial. Terrestrial animals tend to be more vulnerable to predators than are arboreal ones, but the earliest protohominids should not have been more exposed than many of the living terrestrial Old World monkeys and apes. Nor could they have taken an exceptionally wide variety of foods. No matter how mixed their diet may have been, it can hardly have been more diversified than those of many monkeys and some prosimians.

In these circumstances, there would seem to have been no need for them to develop a peculiar communication system simply to cope with familiar features, the same old aspects of the environment of interest to other species. Instead, they probably had to deal with some new or newly relevant matters, of lesser or different significance to their relatives and competitors. There is some evidence to support this hypothesis. The actual foods and other resources utilized by the ancestors of man may have been conventional, but some of the methods by which they were obtained and exploited probably were not. It is almost certain that they involved or entailed the regular and frequent use of tools, more tools and a greater assortment of them than are (or have been) used by any other animals.

Such tool-using must have developed at a more or less early stage in the evolution of the human lineage and adaptive complex. It may, in fact, have been the original "invention" that predetermined everything else. The fossil form that is supposed to have been the first hominid, *Ramapithecus* of the Miocene, is known from a few jaws and teeth, which differ in only a few features from the corresponding parts of some contemporary apes, *Dryopithecus* spp., one or more of which may have been the direct ancestor(s) of the living great apes. The most significant of the diagnostic features of *Ramapithecus* is a reduction in relative size of the incisor and canine teeth, the beginning of a process that has been carried further in modern man. Incisors and canines are very important to nonhuman primates in fighting and display as well as feeding. Their reduction in *Ramapithecus* would suggest that they may have begun to be supplemented or replaced in some of their functions. If so, the replacements can hardly have been anything but tools and weapons (tools for fighting), just as in man. See Hutchinson (1963) and Simons and Pilbeam (1965).

It has also been suggested (by C. M. Jolly and others, see discussion and references in Simons, 1972) that the dental characters of *Ramapithecus* were adaptations to a diet of small seeds, grains, and other vegetable material gleaned from the ground. This may be (further) evidence that the transition from tree-dwelling to ground-living was precocious. It is highly pertinent, therefore, that Alcock (1972) has shown that the habit of using tools is most likely to arise with the invasion of new habitats and environments. See also Chapter 7.

One can think of several ways in which the use of tools could encourage the use of learned rather than "innate" signals.

All displays, both definite releasers and less definite "merely" informative patterns, and probably all unritualized signals, of nonhuman primates seem to be primarily "subjective" expressions of motivation or emotion, various kinds and combinations of drives or tendencies. On occasion and in a few circumstances, some of them may also function as if they were nouns, names of things, or descriptive terms. Thus, for instance, a monkey that utters an alarm call certainly is expressing fear or a tendency to escape. It is saying "I am afraid." It may also be saying, in effect if not in intention, "There is a predator present." If different alarm calls express different strengths of motivation, as is usually the case, then any single performance might even specify the nature and distance of the predator. At least, a listener should be able to deduce a great deal about the predator and its position from the sound of a sequence of calls.

This is all very well. But the things named and described by living monkeys and apes are not particularly numerous or varied. As soon as the ancestors of man began to use tools regularly and frequently, or to use many different kinds of tools, they must have had to cope with greatly increased numbers of things or classes of objects. When and if the efficient use of tools depended upon communication and cooperation among individuals (and many of the earliest tools may have been employed in group defense or perhaps communal hunting drives), much of the information transmitted must have had to become more detailed and objective. A message such as "I am afraid—there is a predator present—run for your lives!" is sufficient in itself. The proper response is obvious and simple, and the reactions of different individuals do not have to be closely coordinated. Even a less urgent or precise message, perhaps conveying more complex information of longer term interest, such as "I am coming into reproductive mood—get ready!" is adequate without further comment by the signaller. Responses can be stereotyped (although possibly in different ways in different situations). But a message that says "There is a tool present—use it!" is not necessarily very helpful. More often than not, it should be supplemented by a specific proposal for treatment or manipulation of the object in question.

(A note in passing. The evolutionary consequences of predatory habits depend upon the techniques employed. Most gregarious mammals that hunt cooperatively, e.g. Lions, Spotted Hyenas, Wolves, and the African *Lycaon*, have communication systems of the usual animal type. See van Lawick-Goodall and van Lawick, 1970, Kruuk, 1972, Schaller, 1972, and Ewer, op. cit. Thus, if hunting encouraged verbalization at some stage of human history, it probably did so only when and as it also entailed the use of tools.)

The development of tool-using in any context must have been accompanied by strong selection pressure in favor of increasing the number of signals to match the increased number of objects to be signified.

The pressure must have been difficult to satisfy. As indicated above, ritualized displays probably cannot be multiplied beyond a certain point. There seems to be an upper limit to the total number of qualitatively distinct major displays possible. It has been suggested (again in Moynihan, 1970b) that this limitation may be designed to prevent confusion. An increase in number of major displays probably would entail greater diversity of forms of displays. It probably also would mean that some displays could be given only rarely. Rare displays of peculiar form might be "unexpected" too frequently in circumstances in which surprise would be disadvantageous.

This suggestion still seems plausible. It will have been noticed, however, that both unritualized "innate" and learned signals should have the same drawbacks as displays in similar circumstances. They also should tend to be surprising when rare, and some of them must become relatively rare when and as they become more numerous and diversified. The fact that learned signals were selected for during the evolution of man would imply that they must have some added value or compensatory advantage, some features not shared with other kinds of signals, in the context of tool-using. I think that this can only be the ease and speed with which they can be *both* acquired and lost.

Consider the probable historical sequence. At some stage, the ancestors of man began to use tools. Their first tools probably were *objets trouvés*, sticks, stones, branches, animal bones, straws, etc., found lying about. (There is a big difference, and almost certainly a large time gap, between using and making tools. The first tools obviously fashioned by man date back to around the beginning of the Pleistocene. They must have had antecedents and precursors.) Objects that can be used as tools without modification usually are rather scattered. Different individuals and groups of early hominids must have found different kinds and combinations of suitable objects available in different areas. They must, therefore, have had to be thoroughly eclectic in their choice and handling of them. Some objects may have been used in very different ways or very rarely.

It probably would not have been efficient, even if feasible, to develop "innate" signals to identify all the objects that could be used and to specify the methods by which they could be manipulated, alone or in combination. The possible numbers and permutations would have been prohibitively large. More important, "innate" signals can be developed only by natural selection of random mutations and genetic recombinations, by evolution over several generations. They cannot be produced "on call" to permit immediate exploitation of sudden opportunities or nonrecurring resources. Yet it is just this faculty that would have been most advantageous to the ancestors of man. Thus, one can see that the principal selection pressure upon protohominids would not have been in favor of elaborating more and more "innate" signals, each one adapted to a particular object or relationship, but rather for a certain kind of inventiveness, a capacity to produce and to react to new signals on an ad hoc and experimental basis within a few seconds or minutes whenever a new situation was encountered. This capacity could only be learning.

The evolution of flexibility or inventiveness probably took as long as that of any "innate" signal, if not longer. Once it was completed, however, then further and learned changes could occur at an accelerated rate, while every new "innate" signal that might have started to develop would still have to go through the whole lengthy process of natural selection, just like all the others that had gone before. Learned languages may be conservative in some groups and societies, but almost never as much so as other kinds of communication systems.

An additional point that may have been of equal or greater importance is the ability to discard and disregard signals. "Innate" patterns are difficult to suppress (or repress) entirely during the lifetime of a single individual. But many learned patterns are easy to forget. If the ancestors of man really did encounter very different assortments of potential tools in different areas, then it may have been advantageous for them to have been able to drop the habits acquired at one site as soon as they moved on to another.

Once learned signals began to be applied to tools, they also became available for communication of more abstract information. It seems likely, however, that they began to be used for abstraction only slowly and perhaps after a long delay.

The hypothesis that reliance upon learned signals is a consequence of reliance upon tools has an aesthetic attraction. It would explain the most peculiar feature of the social behavior of man as an effect of the most peculiar feature of his ecology. Rather less directly, it might also help to account for some puzzling differences and failures among other species.

Most of the monkeys that have paralleled man in adopting terrestrial habits, the baboons and some of the macaques, seem to differ from him in preferring another channel of communication. Not only are their signals largely "innate," but the majority of them are supposed to be visual rather than acoustic (see, for instance, Cole, 1963). It is quite possible that visual communication has special advantages in the open environments with which terrestrial habits are often correlated. But the advantages may not have been accessible to man or his immediate ancestors. An elaborate repertory of visual signals probably would be inappropriate or nonfeasible for an animal that used tools regularly and frequently, simply because its hands and arms and other parts of the body would be preoccupied in managing the tools themselves. Thus, the use of tools may be one of the reasons why most of man's signals are vocal. Conversely, the use of visual signals could be a reason why macaques and baboons do not usually employ tools under natural conditions (although they may do so in captivity—see, for instance, Beck, 1973).

The semiterrestrial chimpanzee, *Pan troglodytes*, is an even more striking case. Individuals of this species do use found tools occasionally, not very frequently but apparently regularly, in the wild (Kortlandt, 1967 and 1974, van Lawick-Goodall, 1970, and Sabater Pi, 1974). A few have been trained, with difficulty and patience, to use a learned and artificial gestural or visual language in captivity (the Gardners, 1969 and 1971, Premack, 1971, Premack and Premack, 1972, Rumbaugh et al., 1973), even though the normal communication system of the species, in its homeland, would appear to be as largely "innate" as those of most other animals. Again there may have been a conflict for possession of the executive organs. Chimpanzees often use their hands for locomotion, "knuckle walking" on the ground or brachiating in the trees.

These facts would seem to argue against the theory of Hewes, 1973, that the verbal languages of man somehow have been derived from or have replaced an earlier system of gestural communication. Of course, the ancestors of man, like all other primates, must have had visual signals. They must also have had acoustic signals. In the present state of our knowledge, it is simplest to suppose that the verbal languages of man are direct descendants of the vocalizations of his ancestors.

The capuchins of the New World may be caught in the same dilemma as baboons and chimpanzees. It is perhaps surprising that they have not become more manlike, since they manipulate objects relatively frequently and have evolved the capacity to perform patterns that could be used as learned signals. They may have been unable to exploit these talents as fully or rapidly as might be desirable because many of the potential signals, many quirks, involve the hands and arms. Possibly their further progress has been, or will be, blocked by this confusion or overlap. Possibly not. It would be interesting to see what happens to them in the next million years or so, in the improbable event that they should be allowed to survive so long.

REVIEW OF CHAPTERS 1-5

The characteristics of New World primates might be summarized as follows:

1. The systematic group as a whole, the family Cebidae in the broad sense, includes a substantial number of species. They can be assigned to some twelve or thirteen genera, divided among eight or nine subgroups. The subgroups seem to have evolved independently from a common ancestor. Their development has been divergent in most respects, but parallel or even convergent in others.

2. At the present time, there is seldom more than one species per genus in any given area.

3. The various species, genera, and subgroups are diverse in some aspects of behavior and ecology. They occupy and exploit different habitats in different ways.

4. They are all, nevertheless, more or less arboreal, inhabiting forest and scrub. Only a few species come down to the ground regularly. None has moved into open country permanently.

5. One genus and species, *Aotus trivirgatus*, is nocturnal. The rest are diurnal.

6. The various forms range from very small to medium large in size. The majority are on the small side.

7. Modes of locomotion include vertical clinging and leaping, quadrupedal running and springing, and brachiation. Prehensility of the tail has been evolved by several genera, presumably independently, in at least three of the subgroups.

8. Different forms feed on different things. The preferred foods are small animals, usually arthropods, and fruits of various sizes, shapes, structures, and degrees of hardness. Only

a few species, most notably the howlers, concentrate upon leaves.

9. Several forms have evolved highly specialized hands, of one sort or another, to facilitate arboreal locomotion. Only one genus, *Cebus*, has developed considerably increased ability to manipulate objects.

10. Different species occur in different social arrangements. There seem to be two basic types of intraspecific organizations. One is the nuclear family, sometimes extended. The other is the large troop, usually or often including subdivisions, temporary sexual partners, families, or age or sex classes.

11. There are some, but not many, obvious correlations between basic social types and abundance of the monkeys themselves, densities of populations, activity rhythms, or crude food preferences (for animal or vegetable materials). Other and more subtle correlations may have escaped detection. It also seems likely, however, that some New World primates have acquired a certain amount of "social independence" from their environments; that they really can, for instance, combine different social relations with similar feeding habits and different feeding habits with similar social relations.

12. Serious aggression and contact fighting are not common in most species. Neither are sexual or sexually derived patterns, apart from actual reproduction.

13. Interspecific social reactions tend to be moderately complex. They can be friendly or unfriendly, and subserve different functions at the same or different times. They may involve other monkeys, squirrels, and birds.

14. Most social relations are partly mediated by ritualized and largely innate signals or displays. All species have roughly comparable numbers of kinds of displays, irrespective of their social organizations.

15. Individuals of several species may develop elaborate aberrant behavior patterns, "quirks," in unusual circumstances. These patterns could provide a basis for learned language, but apparently do not do so to any significant extent.

16. The various forms differ in relative brain size and intellectual development. Greatly increased intelligence seems to have evolved independently several times, in two or three lineages (*Lagothrix-Ateles, Cebus,* and possibly some species of *Pithecia* s.l.).

chapter six
COMPARABLE RADIATIONS

There are only two or three other groups of modern primates that have radiated on something like the same scale as the ceboids. These are the lemuroids of Madagascar and the monkeys and apes of Africa and Asia. A brief comparison of the various radiations may reveal some further aspects of the evolution of primates in general.

The Malagasy lemuroids are the easiest to describe. All the species of the group probably are quite closely related to one another. It is only unfortunate that their fossil record is brief, nonexistent before the Pleistocene or sub-Recent. The known fossil forms indicate that the surviving species are only a partial selection of the range of types that flourished a few hundreds or thousands of years ago.

There is more information on the Old World monkeys and apes, including some Tertiary genera, but it is rather less digestible. All the cercopithecoids and hominoids are often lumped together in a separate taxon of major rank, an infraorder Catarrhini, in contradistinction to the ceboids (usually called Platyrrhini in this context). The precise nature of the phylogenetic relationship between the two groups of catarrhines is not yet altogether clear (see, for instance, Simpson, op. cit., and Simons, 1972), but they certainly have evolved in the same places and probably at the same times, presumably in response to the same or similar pressures.

Lemuroids of Madagascar

The notes on these animals in Chapter 1 may be expanded by more details.

The best accounts of the behavior and ecology of living lemuroids are by Petter (1962a, 1962b, 1965, and 1972), Petter and Pariente (1971), A. Jolly (1966), Martin (1972), Hladik and Charles-Dominique (1971), and Richard (1974). Andrew (1963) describes and analyzes the signal patterns of several species in captivity. Hill (1953) and Remane (1956) list the fossil forms. There is a more extensive discussion of the inferred ecology and habitus of recently extinct species in Walker (1967a, 1967b, and 1974).

A conventional classification would distinguish seven subgroups at the family or subfamily levels and approximately eighteen genera. These are: the subfamily Cheirogaleinae of the family Lemuridae (with the living genera *Microcebus, Cheirogaleus, Allocebus,* and *Phaner*); the subfamily Lemurinae of the Lemuridae (the living *Lepilemur, Hapalemur, Lemur* itself, and *Varecia*); the subfamily Indriinae of the family Indriidae (with the living *Avahi, Propithecus,* and *Indri,* and the extinct *Palaeopropithecus, Mesopropithecus,* and *Archaeoindris*); the subfamily Hadropithecinae of the Indriidae (only the extinct *Hadropithecus*); the subfamily Archaeolemurinae of the Indriidae (only the extinct *Archaeolemur*); the family Daubentoniidae (the barely surviving *Daubentonia*); and the family Megaladapidae (only the extinct *Megaladapis*). Possibly this classification is oversplit. It is evident, nevertheless, that the lemuroids are or were a group of considerable structural and systematic diversity.

Interestingly enough, the recognized systematic units do not always coincide exactly with the major ecological and behaviorial categories, which follow.

1. A series of species or well marked subspecies of the Cheirogaleinae. There are small to very small in size, thoroughly nocturnal, with much the same coloration as other animals of similar habits, largely buffy or grayish above and lighter below. All the forms that have been observed in the field are essentially arboreal, quadrupedal, and move along or among small or medium branches. Most of them eat a variety of foods, many kinds of small insects and fruits (or even flowers). *Microcebus murinus* may be preferentially insectivorous. *Cheirogaleus* would appear to be more frugivorous. Some species

become torpid under favorable conditions. Forms of *Microcebus* and *Cheirogaleus* build nests, often in holes in trees, in which they sleep during the daytime and when torpid. All the species are almost completely nongregarious when active (although there may be concentrations of individuals in nests, at least among *Microcebus murinus*). The most conspicuous social signals are vocal. Most of the sounds uttered are high pitched. Many species are separated geographically or by habitat. Apparently there are few or no overlaps among species of the same genus, but some species of different genera are sympatric over wide areas. None of these forms is known to have developed specialized interspecific reactions to any other.

2. *Lepilemur* and *Avahi*. These also are nocturnal, arboreal, not very large, and dull colored. Neither is very gregarious. *Avahi* individuals usually occur in small family groups. Both types are vertical clingers and leapers. *Lepilemur* eats leaves, bark, and fruit. *Avahi* may be even more strongly folivorous. As would be expected, both genera seem to rely upon vocalizations to a large extent and have few remarkable or complex interspecific reactions.

3. The Aye-aye, *Daubentonia*, is yet another nocturnal, arboreal, and nongregarious type. It is of medium size, with long grizzled blackish hair (reminiscent of some *Pithecia*), and is basically quadrupedal. It also has some extreme morphological specializations: rodentlike, gnawing front teeth and elongated, attenuated, third fingers on each hand. It eats both vegetable and animal foods. The teeth permit it to gnaw into very hard fruits, even coconuts(!), and the peculiar fingers are used to probe into trunks and branches for grubs. Perhaps the Aye-aye is similar in ecology to some of the primitive rodentlike primates of the Palaeocene or one of the marsupial phalangers of New Guinea, *Dactylopsila*, which has equally elongated fingers (in this case the fourth rather than the third). It can hardly be in serious competition with any of its contemporaries in Madagascar. It builds large sleeping nests in trees, not in holes, like some squirrels and apes. In its present

reduced and precarious state, it seems to have become a commensal of man. Almost the only suitable trees left in its range are in and around native villages.

4. The "true" lemurs, *Hapalemur, Lemur,* and *Varecia* are more monkeylike than any of the preceding forms but less so than some of the Indriidae. They are quadrupedal, more or less arboreal, diurnal, or crepuscular. Some have complex social relations.

Hapalemur griseus (the only species of the genus that has been observed frequently) is not entirely typical. It shows points of resemblance to the Cheirogaleinae. It is medium sized, dull colored, and occasionally active at night. Individuals usually live in not very large family groups. *Lemur* seems to include four species: the highly polymorphic *macaco*, the more uniform *mungoz* and *rubriventer*, and the rather distinctive *catta*. All are moderately large in size. The first three species are almost completely arboreal. The last comes down to the ground from time to time, perhaps even more frequently than *Cebus capucinus*. All of them live in large social groups, real troops that may contain several adults of both sexes in addition to juveniles and infants. Individuals of *Varecia variegatus* (probably the only surviving species of the genus) are larger than any *Lemur* in size, but resemble *Hapalemur* in living in smaller social groups and sometimes moving about at night. All forms of all three genera seem to be primarily or exclusively vegetarian. *Hapalemur griseus* may be as folivorous as *Lepilemur*. *Lemur* and *Varecia* eat a wider assortment of plant foods. *Varecia* and *Hapalemur* build crude nests for their young. All the species utter many and diverse calls and notes. *Lemur catta* and *Hapalemur griseus* also have elaborate olfactory signals. All forms of *Varecia variegatus, Lemur catta,* and some forms of *L. macaco* are brightly colored. Many of them tend to emphasize black, white, or rufous.

Some of these species overlap very broadly. There are no records of special interspecific reactions among them under completely natural conditions, but Jolly has described how a single *Lemur macaco* was able to insert itself into, and play a significant social role within, a group of *L. catta* in an overgrown tamarind plantation.

Mesopropithecus seems to have been another arboreal and diurnal springer and may have been similar to the "true" lemurs in way of life.

5. The Indri, *Indri*, and the sifakas, *Propithecus*. These are larger than the other surviving lemuroids and more anthropoid in shape and proportions. They seem to be completely diurnal, at least when not disturbed. They spend most of their time in trees, and they are primarily vertical clingers and leapers. *Indri* may also brachiate occasionally. Individuals of some or all forms come down to the ground in certain circumstances. On the ground, they are supposed to leap bipedally. (Some of the structural concomitants of indriid modes of locomotion are discussed by Stern and Oxnard, 1973.) All species eat leaves, fruits, and bark. The best known species, *Propithecus verreauxi*, lives in varied and flexible social groups, perhaps most often in small bands or extended families. All forms have many ritualized signals. *Indri* itself is famous for the loudness and complexity of its operatic vocalizations. Most forms have conspicuous color patterns. They show the maximum refinement of the diurnal lemuroid tendency toward bizarre or intricate arrangements of white, black, brown, and orange rufous. They probably are ecologically equivalent to such vegetarian monkeys and apes as *Alouatta, Colobus*, or *Hylobates*.

6. Other offshoots of the same stock seem to have carried convergence even further. *Palaeopropithecus* would appear to have been a large, brachiating or "hanging," arboreal, diurnal, and vegetarian animal, like the living Orang-utan (*Pongo*).

7. *Archaeolemur* and *Hadropithecus*, of moderately large size, differ from one another in some details of anatomy, but both probably were diurnal quadrupeds living on the ground in open country. The teeth of the first genus would seem to indicate that it was adapted to feeding on grasses, grass corms, and grains. The latter may have fed on large tough objects

or small items (seeds, etc.) gleaned from the surface of the ground. Either one or both may have been equivalent to the baboons *Papio* and *Theropithecus*.

As mentioned in Chapter 1, the brains of living lemuroids are relatively smaller and simpler than those of monkeys and apes. The brains of such genera as *Lemur, Propithecus,* and *Indri* may be somewhat more highly developed than are those of the Cheirogaleinae. *Archaeolemur* and *Hadropithecus* had larger and more rounded braincases than do the surviving lemuroids. It is possible, therefore, that they were also more intelligent. The analysis of casts of braincases by Radinsky (op. cit.) would not seem to support this suggestion, but does not definitely disprove it.

8. *Megaladapis* was gigantic, perhaps comparable to another marsupial, the Australian Koala (*Phascolarctos*), but much larger. It may have been an inflated teddy bear, diurnal, arboreal, a vertical climber, slow moving, stupid, and vegetarian (see also Zapfe, 1963). The almost equally large *Archaeoindris* may have had similar habits.

Some features of this assortment of lemuroids are worthy of note. They include diet and activity rhythms as well as brain development.

They should be viewed in geographical and ecological perspective. Madagascar is a very large island, but of simple structure. It is essentially a single plateau, with only a few scattered high peaks, surrounded by more or less extensive coastal lowlands. It is entirely tropical or subtropical. There is considerable variation in rainfall. The eastern coast and escarpment are very humid; the western slope is less humid; and the extreme south is very dry. The native vegetation ranges from rain forest to semidesert thicket of Didieraceae and *Euphorbia*. Most of the center is degraded grassland now, but the spread of this grassland would appear to have been very recent. The island has been separated from other land masses for a very long time, probably since the late Mesozoic. The fauna reflects this isolation. The birds are only moderately diverse and the terrestrial mammals belong to only a few taxonomic groups. Some of these (the large and flightless "elephant birds"—the "rocs" of Sindbad the Sailor, the fly-catcherlike "vangid" shrikes, cuckoos of the genus *Coua*, tenerecine insectivores, cricetid rodents, and viverrid carnivores) underwent modest radiations of their own. This fauna is not, perhaps, really impoverished, but it is much less rich than are those of continental Africa, tropical Asia, or Central and South America. (There are descriptions of the relevant aspects of Malagasy history and conditions in Paulian, 1961, and Battistini and Richard-Vindard, 1972. See also Rand, 1936, and Moreau, op. cit.)

The development of brains and intelligence will be discussed again later. Only a few points need to be mentioned here. Even if *Archaeolemur* and *Hadropithecus* were progressive, it would still be true that the lemuroid radiation produced relatively few types as intelligent as some of the ceboids ("*Cacajao*," *Lagothrix, Ateles, Cebus*) and most of the African and Asiatic monkeys and apes, and that the Malagasy primates would seem to have had to cope with a less varied array of competitors of other groups, predators, and (possibly) foods.

It is remarkable that so many lemuroids are or were vegetarian and so few were omnivorous or insectivorous. This may be partly due to a distinctive division of niches. (Lemuroids may have had to meet fewer kinds of competitors than monkeys and apes. This does not mean that they had no competitors, or that actual or potential competitors may not have been numerous as individuals.) Interactions with birds may have been a determining factor. Moreau notes that Madagascar is comparatively deficient in fruit- and seed-eating birds, while the endemic vangid shrikes are almost all insectivorous. At the present time, at least, there seems to be a more clear-cut ecological distinction between primates and birds in Madagascar, less overlap, than between their relatives in other and more continental parts of the tropics. It is tempting to suggest that the simplicity of the division in Madagascar is somehow due to the lesser number of species involved.

The lemuroids are also characterized by a high percentage of nocturnal forms, both at the generic and specific levels. Many or all of the recently extinct types may have been diurnal, which would not affect the validity of the generalization as a whole. The proportion of nocturnal types may have been less important during the Pleistocene than at present. It must, nevertheless, have been higher than in some other regions. Aside from *Aotus* and perhaps *Cercopithecus hamlyni*, there are no living monkeys or apes that are really nocturnal. Where nocturnal primates of other groups do exist, as in Africa with its pottos and galagos, they are less diverse than the nocturnal lemuroids. The fossil record, such as it is, would suggest that this has been true in Africa since the middle Tertiary.

There seems to have been a positive proliferation of nocturnal types in Madagascar. The nocturnal lemuroids probably occupy habitats or niches that are filled by animals other than primates elsewhere. Nocturnal primates may be restricted in Africa and Asia by competition with rodents, murids, sciurids, and others. Certainly, the equivalent niches in tropical America are largely filled by opossums of the family Didelphidae and, to a lesser extent, raccoonlike Carnivora.

There also seems to have been a negative factor, a nonproliferation of diurnal types. This is not conspicuous at the generic level. Including the extinct forms, the Malagasy primates have nearly as many definitely, probably, or largely diurnal genera as either of the other two major radiations. (The correspondence is remarkably close. Using the rather conservative classification outlined in Chapter 1, there are 12 genera of living diurnal ceboids on the mainland of tropical America. According to the less conservative classification of lemuroids cited immediately above, there were 11 genera of diurnal primates in Madagascar in the sub-Recent. There are approximately 12 genera of living diurnal primates in tropical Africa.) One might perhaps be surprised that the diurnal types of Madagascar have not expanded to fill more niches like their nocturnal counterparts. Be that as it may, the living diurnal lemuroids do seem to have relatively few species per genus, fewer species (and less sympatry among them) than either their nocturnal relatives or even the ceboids. This must (again) be correlated with fewer opportunities for speciation in isolation. The structure of Madagascar is even simpler than that of South America. The diurnal lemuroids are larger than the nocturnal forms. Diurnal individuals may, therefore, range over wider areas.

It would be interesting to consider some of the possible advantages and disadvantages of nocturnality versus diurnality in primates. There is good evidence that the ultimate ancestors of all placental and marsupial mammals were nocturnal (see, for instance, Walls, 1942). The diurnality of many primates must be a derived condition.

Why should it have been selected for? Probably not to help in feeding. There is almost every conceivable kind of alimentary regime among both diurnal and nocturnal species of mammals and other terrestrial vertebrates as a whole. The obvious alternative is some aspect, consequence, or correlate, of predation.

Diurnality may provide protection. Some other comparative data are suggestive in this connection. The phalangerid marsupials of Australia and New Guinea, in addition to *Dactylopsila* and *Phascolarctos*, include a whole series of less exaggerated arboreal and frugivorous, omnivorous, and/or preferentially insectivorous types of various sizes (a total of 17 genera according to Keast, 1972c). Many of them are ecologically equivalent to primates in many respects, but they are almost all nocturnal (Tyndale-Biscoe, 1973).

Predatory birds, both diurnal hawks, falcons, and eagles, and mostly nocturnal owls, are more or less abundant and similar in behavior and ecological adaptations everywhere that primates and primatelike animals occur. Carnivorous mammals are more heterogeneous. All the predatory mammals of Madagascar,

Africa, and Asia are placental. So are almost all of the modern South American forms (the large semiomnivorous opossums of the genus *Didelphis* may be the only important exceptions—they certainly will take any prey that they can get). The great majority of the predatory mammals of Australia and New Guinea, on the other hand, are marsupial, members of the family Dasyuridae.

The viverrids of Madagascar, moderately diversified as they may be, and despite their placental status, are less varied and less specialized than the Carnivora of Africa, Asia, and the Americas (see Albignac, 1973). Perhaps only one of the Malagasy forms, *Cryptoprocta ferox*, can catch lemurs. The dasyurids, for whatever reason, seem to be comparatively unintelligent and even more unenterprising. They have not been able to compete successfully with the placentals recently introduced into Australia by man.

There might, therefore, be a causal relation between the efficiency or effectiveness of predatory mammals and the activity rhythms of their potential prey. Predators can be most dangerous during either the periods of movement or the periods of repose of the prey. A monkey or monkeylike animal stands a good chance of escaping from a mammalian predator as long as it is awake and alert. It is much more vulnerable when asleep. Predatory mammals probably are more dangerous to sleeping prey than are predatory birds because they hunt by smell as well as sight and sound (most birds have little sense of smell) and are more likely to investigate sleeping refuges such as nests and holes in tree trunks. A suddenly awakened and startled animal probably has a better chance of escaping in the dark than in the light. Thus, one might expect that the greater the effectiveness of the local carnivorous mammals, the stronger will be the selection pressure in favor of sleeping at night rather than during the day.

What this is leading up to is the suggestion that most primates have become active during the day in order to be able to sleep in comparative safety at night.

Monkeys and Apes of the Old World

Many species of Catarrhini have been observed by many students in the field and in the laboratory. The books and papers produced as a result are almost literally countless. I shall not (could not) attempt to list even a small fraction of them. It may be useful, however, to cite some works that appear to be representative of ecological and behavioral descriptions and analyses of comparative interest.

There is information on single species and small groups of closely related species in Altmann (1962, 1965, 1968a, 1968b), Altmann and Altmann (1970), Angst (1974), Bertrand (1969), I. Bernstein (1967b and 1968), A. H. Booth (1957), Bourlière, Bertrand, and Hunkeler (1969), Buxton (1952), Carpenter (1940), Chalmers (1968a, 1968b, 1968c), Chivers (1974), Crook (1966), Crook and Aldrich-Blake (1968), Deag and Crook (1971), Dunbar and Dunbar (1974), Gauthier (1967), Gauthier-Hion (1966), Haddow (1952), Hall (1965), Hall and Gartlan (1965), Hunkeler, Bourlière, and Bertrand (1972), Imanishi and Altmann (1965), Jones and Sabater Pi (1968), Kummer (1968), van Lawick-Goodall (1968), MacKinnon (1974), Marler (1970, 1972), Munemi and Tetsuzo (1972), Nolte (1955), Poirier (1969, 1970), Rahaman and Parthasarathy (1969), Rodman (1973), Rowell (1973), Schaller (1963), Struhsaker (1967), Sugiyama (1965 and 1973), Tenaza and Hamilton (1971), and Ulrich (1961).

General surveys or collections of accounts of many different species can be found in Altmann (1967), Andrew (1963 and 1964), I. Bernstein (1967a), C. Booth (1962), Buettner-Janusch (1962), DeVore (1965), Dolhinow (1972), Gauthier and Gauthier-Hion (1969), C. M. Hladik and A. Hladik (1967 and 1972), Jay (1968), A. Jolly (1972), Michael and Crook (1973), Napier and Napier (1967 and 1970), Poirier (1972), Reynolds (1967), Rowell (1972), Struhsaker (1969), Tappen (1960 and 1968), Tuttle (1972), and Van Hoof (1962).

There are bibliographies in Baldwin and Teleki (1972, 1973,

and 1974), and additional relevant material in almost every issue of the journals *Primates* and *Folia Primatologica* and many periodicals concerned with ethology or anthropology.

All the living and Pleistocene Old World monkeys can be placed in a single family Cercopithecidae, with two subfamilies, Cercopithecinae and Colobinae. The Cercopithecinae include two groups of genera. These are: first the guenons (*Cercopithecus* and the barely separable *Miopithecus* and *Allenopithecus*) and the Patas Monkey (*Erythrocebus*), and secondly the mangabeys (*Cercocebus*), macaques (*Macaca*), and baboons (*Papio* and *Theropithecus*). The Colobinae include the guerezas (*Colobus* and perhaps *Procolobus*) and the langurs (*Presbytis* and several other genera, at least *Pygathrix* and *Nasalis*.) As in the lemuroids, the systematic and ecological categories do not always coincide. For purposes of rapid comparison, it may be convenient to group the forms as follows.

1. The guenons and mangabeys. There are a great many kinds of these monkeys; some 19 to 20 species of guenons in something like 11 superspecies, and at least 4 species of mangabeys in 2 superspecies. They are all African and almost entirely restricted to forest and scrub. They are quadrupedal springers and runners, with tails that are either nonprehensile (in the guenons) or only very slightly prehensile (in some of the mangabeys). Some species are entirely arboreal. Others divide their time between the trees and the ground. They are among the typically monkeylike forms, somewhat reminiscent of *Saimiri* and *Cebus* in general shape and proportions. They have a considerable range of sizes but average larger than their New World analogues. Many of the species are brightly or conspicuously colored, often in intricate patterns. Their fur can be olive, yellow, white, gray, brown, rufous, or black. Patches of naked skin from the face or in the genito-anal region can be black, white, blue, green, pink, or red. As a whole, they exploit most of the habitats available. There are numerous and extensive overlaps among species. Competi-

tion is reduced and coexistence permitted by special preferences for different kinds of forest or scrub, levels of vegetation, and particular foods (all or most take both animal and vegetable materials, but the mix seems to be different in each case), and perhaps by different activity rhythms.

Some are real commensals of man. The populations of *Miopithecus talapoin* studied by A. Gauthier-Hion in Gabon may be more specialized in this respect than are any ceboids. They are supposed to feed largely upon tubers of manioc, which are collected and then soaked and stored in wells (under water) by the local people. The habit is all the more remarkable as manioc is of American origin and has been cultivated in Gabon for only a few centuries.

2. *Erythrocebus patas* is a medium large, dull colored, terrestrial, long legged, cursorial guenon of African savannahs.

3. *Macaca* is a genus of 11 to 13 species, widely distributed in southern Asia, extending to many islands and into the subtropical and temperate zones in China, Japan, and the Mahgreb north of the Sahara. The various species are of medium to medium large size, quadrupedal, usually rather heavy in build, with strong and somewhat elongated jaws and more or less abbreviated tails. They have rather uniform and dull colored fur, sometimes with simple ruffs or tufts around the head, and areas of pink, red, or dark skin on the face and in the genito-anal region. They are almost ubiquitous within their range, partly arboreal and partly terrestrial, in many different types of vegetation and in open areas, including such peculiar habitats as beaches and rocky shores. As in the guenons, there are many geographical overlaps among species. Overlapping forms maintain some ecological and social distance by the usual sorts of differential preferences. All or most forms eat many kinds of animal and vegetable foods. The Japanese species, *fuscata*, is known to be an experimentalist, prepared to try new kinds of food when given the opportunity. Both *fuscata* and the common Indian form, *mulatta*, the Rhesus Monkey of the laboratory, have become semidomesticated.

4. Baboons look like exaggerated macaques, larger, longer snouted, shorter tailed. They also are more often terrestrial. The genus *Papio* includes two brightly colored species in the forests of west Africa and one dull colored species or super-species of open and semiopen country, gallery forest and forest edges, in much of the rest of Africa. They are primarily vegetarian, foraging for all sizes of roots, seeds, fruits, leaves, etc., but will take animal foods, arthropods and even small mammals, on occasion. The only living species of *Theropithe-cus,* is a dull colored, ground living, inhabitant of the barren Ethiopian highlands. It feeds on such small items as grass seeds and grains. Extinct forms of the same genus were more widespread in east and south Africa during the Pleistocene. Some were gigantic.

5. The Colobinae have a complex, "sacculated," structure of the stomach. This may facilitate the breakdown and absorp-tion of leafy material. The purely African guerezas do, in fact, seem to be largely or exclusively folivorous. There are at least several species of these animals, broadly overlapping in many areas. They are arboreal, medium sized, quadrupedal, long tailed, and brightly or conspicuously colored. The Asiatic lan-gurs are similar in basic morphology but generally duller in color. Some of them, e.g., *Nasalis larvatus* of Borneo, also seem to depend upon leaves and shoots. The more dominant genus *Presbytis* is more diversified. It occurs throughout south-ern Asia and many adjacent islands. It includes many forms, 14 or more species divided among at least 4 species groups. They have become almost as varied in habits as the species of *Macaca* and *Cercopithecus*. Some are exclusively arboreal, with incipient tendencies toward brachiation. Others come down to the ground regularly. None of the living forms seems to be as frequently terrestrial as some macaques, to say nothing of baboons or *Erythrocebus*. (It is possible, however, that some of their Tertiary ancestors or relatives were inhabitants of open country.) Some forms of *Presbytis* certainly take other

foods in addition to leaves, although they may always be more vegetarian than some of the more adventurous macaques. The Sacred or Hanuman Langur of India, *P. entellus,* is the most famous of the primate commensals of man and has proved to be very flexible and adaptable indeed.

All living cercopithecoids have complex social behavior. All the forms that have been studied are more or less highly gregarious. Their basic intraspecific social units range from large bands with several adults of both sexes (e.g., many guenons) to "one-male groups" or polygynous families (the Patas and some open country baboons), which sometimes merge into larger assemblages or may be supplemented by ancillary units (e.g., troops of bachelor males). Some forms have also developed and elaborated cooperative interspecific bonds (Gauthier-Hion and Gauthier, 1974).

All or most species are rather intelligent. There is no reason to suppose that any one of them is cleverer than *Cebus* or *Ateles,* but none falls far below. None seems to be as relatively simple as many lemuroids, the tamarins and marmosets, *Aotus, Callicebus,* or *Alouatta.*

The living apes comprise the gibbons and Siamang (*Hylobates*), the Orang-utan (*Pongo*), the chimpanzees (*Pan*), and the Gorilla (either a separate genus—logically enough *Gorilla* —or a subgenus of *Pan*).

1. The gibbons and Siamang are the most monkeylike. There are approximately 5 species. They are widespread in forested parts of southeast Asia. They are lightly built but moderately large on the average. Most of them are comparable to the larger Colobinae in size. The exceptions are the Siamang, *syndactylus,* which is appreciably larger and heavier than any of the others, and a dwarf form, *klossii.* They are all arboreal, expert and agile brachiators, often fast moving, vegetarian (frugivorous or folivorous), and not very gregarious. They usually occur in nuclear family groups.

2. The Orang-utan is very different from gibbons in details

of internal anatomy and physiology, but less so in habitus. It also is arboreal, brachiating, and frugivorous (taking some animal food as well). Its most conspicuous distinction is size. It is much larger and heavier than even the Siamang, with a larger, perhaps better developed, brain than any gibbon or monkey. It may, in fact, be as heavy as any arboreal animal can afford to be. As would be expected, therefore, adults tend to move slowly and cautiously. Individuals have been seen to build nests for sleeping at night and resting during the day, presumably more for support than for concealment. The species has been hunted intensively. It has become restricted to parts of Sumatra and Borneo. Known populations are small and scattered. In these circumstances, Orangs are even less social than gibbons. They may be the least gregarious of diurnal primates. There have been observations of single individuals of different sexes and ages, mothers with young, pairs (mostly sexual consorts), and a few groups of more individuals, sometimes with overlapping home ranges. This sort of grouping, or dispersion, might or might not resemble what used to be the normal social organization(s) of the species when conditions were more favorable.

3. The chimpanzees can be divided among two species, *troglodytes* and *paniscus*. They probably are quite closely related to *Pongo* and have similar brains, but they are African and differ in ecology and behavior. They are much the same size as Orangs, but all the forms that have been studied in the field (mostly *troglodytes* ssp.) have been found to be partly terrestrial, almost omnivorous, and highly gregarious. They often occur in large groups that are both varied and variable, perhaps as much like some troops of spider monkeys or baboons as anything else among primates. At the present time, all the forms seem to be largely restricted to forest habitats, but it has been suggested that some of them may have intruded upon more open areas, savannahs with trees, before they were subjected to pressures from man. The suggestion

seems plausible. Some of their behavior patterns could have originated as adaptations to life in semi-open country. The Gorillas, also African, differ from chimpanzees in being larger, actually enormous, and even more terrestrial. The best known form, *beringei*, is an inhabitant of montane forest and scrub. It seems to be purely vegetarian, concentrating upon large and juicy leaves, stalks, and fruits. Most individuals occur in groups of medium size, one or a few adult males with several adult females and young. These groups seem to be tightly knit and very stable. Some other males live singly, apparently for long periods of time. Both chimpanzees and Gorillas make nests like Orangs. As noted in Chapter 5, individual *troglodytes* have been seen to use (not make) simple tools in the wild.

4. An equally or even more specialized ape, *Gigantopithecus*, occurred in the Pleistocene of eastern Asia. As its name implies, it was another large form, as large or larger than the Gorilla. It would also appear to have been herbivorous, but perhaps more a grazer than a browser, more like *Theropithecus* in diet.

5. Man himself has a common ancestry with the living apes. Presumably he began as another medium or medium large ape of semiterrestrial and omnivorous-herbivorous habits.

All living apes have elaborate signal repertories. There is considerable and obviously adaptive variation from species to species. Some individual patterns are strictly homologous with some of those of man and/or cercopithecoids.

The apes are declining. The surviving forms are remnants or relicts. The stock was more varied in the middle Tertiary. Most of the early forms were more monkeylike than any of the living types, even the gibbons. The apes would appear to have been squeezed from two directions in the course of evolution, by their close relative man (and his ancestors and collaterals) on one side, and by their more distant relatives the cercopithecoid monkeys on the other. The fossil record suggests that the major expansion of the cercopithecoids was substantially later in time than the initial diversification of the hominids. And it is the monkeys that have replaced the apes in their original habitus.

The Catarrhini as a whole, monkeys and apes together, are characterized by a series of distinctive features, mostly matters of proportion and frequency. Among these are: 1. A comparatively very great number of overlaps among species. (This is a purely specific phenomenon. There are never as many overlaps among genera of Catarrhini in a limited area, a subsection of a region, as there are among genera of ceboids in such places as the upper Amazon.) 2. A comparatively high proportion of partly or exclusively terrestrial species. 3. An equally high or higher proportion of very gregarious types. 4. A large average size. (The ceboids do not include any animals as large as an ordinary male baboon. The Catarrhini do not include anything as small as tamarins and marmosets.) 5. A high average degree of intelligence.

These features may have a multiplicity of advantages, may have been selected for, for many reasons. It is also conceivable that they are all correlated with geography, i.e., the large and long scale discontinuity of the habitats in which the Catarrhini have evolved.

The floras of different regions of the tropics are composed of different elements, orders, families, genera, and species (see the papers of Richards, and Raven and Axelrod, cited above); but they seem to be essentially uniform in superficial structure in corresponding climates (Leigh, in press). Thus, the differences among primates of different regions are not likely to be *only* adaptations to their immediate physical or floristic surroundings at any given time. Some of the peculiarities of the Malagasy lemuroids can be ascribed to their distinctive assortment of predators and competitors. In the case of the Catarrhini, it may be more significant that the tropical parts of Asia and the adjacent islands and continental Africa are not only enormous but *enormis* in the original sense of the word. The major habitats within them are scattered in a more intricate pattern than are the rather simple zones of Madagascar and South America. Even before man began cutting trees in earnest, the forests of Africa and southern Asia must have been divided into isolated or semidetached blocks by large bodies of water (e.g., the Arabian Sea and its bays, the Persian Gulf, and the Red Sea) and many expanses of open country (ranging from the Sahara and Somali Deserts to mere patches of not very rigorous savannah). The forests themselves must have been variegated by chains of mountains and individual massifs (the topography of Africa is moderately uneven; that of southern Asia is immensely complicated). The scattering of habitats must have increased and decreased repeatedly during the climatic fluctuations of the Pleistocene. As noted in Chapter 2, forests probably were more reduced and widely separated in Africa than in South America during periods of maximum aridity. All this must have provided ideal opportunities for geographic speciation among Old World monkeys and apes. Most African and perhaps many Asiatic forest blocks should have been invaded many times. It is the forested areas that have the greatest numbers of sympatric species now.

Semiterrestrial habits may be particularly advantageous for species of forests that are unstable, ephemeral, or fluctuating in extent or composition. Primates that are not reluctant to come down to the ground can easily leave a forest that is disappearing or has become unfavorable, and cross open ground to look for another. This, in turn, can be a preadaptation to more permanent life in open areas.

It has already been mentioned that animals on the ground may be more vulnerable to predators than otherwise similar forms in trees. Insofar as large size confers a degree of protection against predators—and it must always discourage the smaller carnivores and birds of prey—there will be new or extra selection pressure in favor of an increase in size as soon as a species starts to come down to the ground. The effects will be reinforced by a reduction of earlier pressure in favor of small size to permit easy movements along small or weak branches in trees. Simply because added eyes and ears facilitate detection of predators, there also will be added selection pressure in favor of an increase of gregariousness as soon as predation becomes more frequent or dangerous.

Both greater size and gregariousness might be expected to favor intelligence, for reasons to be discussed in the next chapter.

chapter seven

THE DEVELOPMENT OF INTELLIGENCE

The various species and groups of primates may not be equally clever, but most of them are, and have been for millions of years, brighter than most other mammals or other vertebrates. Marked increases of intelligence have evolved repeatedly, independently, in different kinds of primates in different areas at different times. The obvious question is, why?

It is very unlikely that mutations for greater brain size or intellectual capacity do not occur in all animals. They may occur with some appreciable frequency. They do not, however, always become established in populations. This suggests two further and more precise questions. Why should natural selection for increased intelligence have been stronger in most primates than in other animals? Or, conversely, why should there have been stronger selection pressure against increased intelligence in most other animals than in primates?

As usual, the problem can be attacked by comparative methods. One may hope to discover correlations between greater or lesser degrees of intelligence and other aspects of behavior or ecology. Unfortunately, the correlations cannot be exclusive, single, or simple. Consider some of the more obvious possibilities.

The great majority of living primates are arboreal. The development of arboreal habits was one of the earliest adaptations of the order, perhaps the initial or principal key to its success. Still there are many arboreal mammals of other orders, e.g., some didelphids, dasyurids, squirrels, other rodents, edentates, that are not noticeably more intelligent than closely related terrestrial forms. Furthermore, there is no evidence that completely arboreal monkeys and apes are more intelligent on the average than their relatives that have secondarily become partly or wholly terrestrial. Indeed, quite the contrary is so in particular cases.

The most intelligent primates are diurnal, but many other diurnal animals are not intelligent.

Some of the cleverer primates are highly gregarious. Some of the most nearly solitary species have comparatively small brains. There is great variation among "intermediate" forms. Many of the most extremely gregarious types, e.g., *Lemur, Saimiri, Alouatta, Macaca*, are not more intelligent than some of their less gregarious relatives, e.g., *Propithecus* or *Hylobates*. Other groups of placental mammals, especially Carnivora and ungulates, can be cited to similar effect. Among the Perissodactyla, the brain of the nearly solitary tapirs, *Tapirus*, seems to be as well developed as that of the highly gregarious horses, asses, and zebras, *Equus*. The frequently solitary cats, Felidae, seem to be approximately as intelligent as the dogs and wolves, Canidae, which tend to live in larger or tighter societies. Within the single family Procyonidae, the gregarious Coatis, *Nasua narica*, do not seem to learn much more, or more rapidly, than the nongregarious raccoons (*Procyon* spp.).

Some earlier students of human evolution, e.g., Elliot-Smith (1927), stressed the importance of fine manipulative ability. Some of the most intelligent monkeys and apes have this ability, but others, such as *Ateles* and *Hylobates*, and probably the uakaris, apparently do not.

Leaving aside simplistic theories, it seems evident that all or most cases of increased intelligence among primates must be causally related to many different factors, only some of which are included in the preceding list. To put the matter in the most general possible terms: increased intelligence may be explained as an adaptation to cope with, or as a result of, increased "complexity of the environment," the relevant, not the total, environment, the number of cues or stimuli to which it is possible, necessary, or advantageous to react. Arboreality, diurnality, gregariousness, manipulative ability, must all affect the number of relevant stimuli presented and perceived. They may tend to increase the number in some or all circumstances. So will other features.

Diurnal species can, and obviously do, receive more visual information than nocturnal species. They do not necessarily receive less acoustic, olfactory, or tactile information. The data gathered by the eyes may be more abundant, varied, or precise than those apprehended by the other senses. Diurnality itself may have originated as an adaptation to cope with certain types of particularly efficient predators, but the sheer diversity of predators encountered may be almost as significant, and many other things could be even more important.

Arboreal animals may be confronted with more different types of foods or pathways than the inhabitants of open country. They also live in a partly three-dimensional world and must pay attention to stimuli coming from all directions. (The species of another group, the Cetacea, the dolphins and porpoises and whales, many of which live in large and deep bodies of water, are more intelligent than most other mammals apart from primates. One of their distinctions is that their world is almost perfectly three-dimensional.)

These ideas are not original. The New World primates provide new evidence to support them, some diagrammatic examples of selective responsiveness to different kinds and degrees of diversity.

Among the larger forms that have been studied at length, *Aotus, Alouatta villosa,* and *Callicebus moloch* appear to be the least intelligent. *Aotus* is nocturnal. *Callicebus moloch* seems to be partly immune to predation and may have few serious competitors, at least in swamp forests. It ignores most other species of vertebrates most of the time. *Alouatta villosa* eats a comparatively restricted number of foods. Presumably it recognizes them by a restricted number of cues.

Some other large forms, the Common Spider Monkey and species of *Cebus,* are known to be particularly intelligent. They eat many kinds of foods. The capuchins also find and manipulate their foods in complex ways. Some or all populations of the Common Spider Monkey have elaborate and unusually variable intraspecific relations. Both spider monkeys and capuchins associate with species of other genera and other families.

An important point suggested by studies of ceboids is that different classes of stimuli can be essentially equivalent. The complexity of intraspecific reactions of *Ateles* would seem to have facilitated much the same increase of intelligence as the complexity of feeding behavior of *Cebus*. The simplicity of social bonds in *Callicebus moloch* would seem to have had as depressing or negative effects upon intelligence as the rarity of visual stimuli in *Aotus*.

Of course, some of these terms are vague. It might be expected that the intellectual faculties selected to favor certain social reactions would be different in quality from the faculties developed to further the discovery and manipulation of certain foods. Perhaps they are; but the distinction is not always very evident in actual fact. Comparable selection pressures can produce remarkably similar results. Capuchins and spider monkeys really do seem to be equally bright in many of the same ways. They even have similar temperaments (revealed, more or less naturally, by the closeness of the associations between them on Barro Colorado Island). Observations of captive individuals (Williams, 1969), suggest that Woolly Monkeys of the genus *Lagothrix*, which are also adaptable and opportunistic but must have a different tempo in the wild, resemble both spider monkeys and capuchins in character as well as intelligence.

The argument may be extended to help explain some (more) of the common or average differences among major systematic groups, Malagasy lemuroids, ceboids, and Catarrhini, in functional terms. All the surviving and all or most of the recently extinct lemuroids appear to be or have been less intelligent than many or most of the modern ceboids and Catarrhini. The ceboids include some forms that are less intelligent than any of the Old World monkeys and apes. There is no reason to suppose that the three stocks ever differed in basic potential; but they have evolved in different circumstances. The lemurs of Madagascar proliferated in an isolated area that is, and presumably always was, less diverse, biologically, than the continental regions of the tropics. Probably, the environments of the Catarrhini were not more diverse than those of the ceboids in general or in most respects; but some of their elements may have been more often obtrusive or "demanding." The wealth of contacts and overlaps among species of the same genus in the Catarrhini probably means that many of them have had to confront more competitors of the most damaging and insidious sort, i.e., rivals with capabilities very similar to their own. Successful competition with such rivals may require more refined attention to more details than does competition with more distant or different rivals—which are likely to be less immediately threatening—or the search for food or escape from predators or any other kind of interspecific behavior. Doubtless the temporal orders or sequences in which sets of relevant stimuli are encountered can also be important.

Several groups of birds show the same evolutionary trends as primates, especially the parrots, order Psittaciformes, and members of the family Corvidae, the ravens, crows, magpies, and jays, order Passeriformes. The parrots appear to be almost exact analogues of monkeys. They are primarily tropical, arboreal, diurnal, characteristic of forested regions, moderately large (for birds) on the average, often highly gregarious, with complex systems of acoustic, visual and other signals, usually brightly colored in intricate patterns, omnivorous or at least very catholic in their choices of foods, and sometimes good manipulators of objects. Many of the larger species also are more intelligent than most other birds (K. Lorenz, 1952, and Rensch, 1956 and 1960). The jays and magpies, the most primitive Corvidae, share many of the same features.

The more specialized crows and ravens may be compared with another primate. Most of them differ from jays and magpies in being larger, preferring more open habitats, being rather more frequently terrestrial (and sometimes scavenging or predatory), and having even better brains and greater intelligence. Some of them also are much more abundant. In other words, they differ from their relatives in some of the ways that protohominids or primitive man may have differed from monkeys and conventional apes.

These examples and others (perhaps even *Archaeolemur* and *Hadropitheous*), indicate that species that have developed a fair amount of intelligence to manage the complexities of diurnal, omnivorous, and gregarious life in forests may be able to invade more open habitats with unusual success, and that their invasions may tend to encourage further increases of intelligence. Species that move into open country by this route seem to be more likely to become more intelligent than other species that have always lived there or that move into it from different habitats or after evolving different ways of life. Probably because they have already acquired the ability to use a great variety of cues and signals, they seem to react to more diverse stimuli in their new surroundings than do the other inhabitants of the same environment. (In the case of man, of course, the number of new cues must have gone up like a rocket with the development of tool-using and learned language.)

It may be mentioned that other groups of mammals that share some characteristics with typical monkeys and apes, but have not evolved the same degree of intelligence, are more limited in their activities in one way or another. They interact with, or are exposed to, less varied stimuli. Thus, for instance, opossums and phalangers, some arboreal rodents and edentates, like *Aotus* and the Cheirogaleinae, are nocturnal or almost completely nongregarious. Coatis tend to rely upon one sense, the olfactory, to a very large extent. The great majority of the more gregarious ungulates do not need to choose among or manipulate any appreciable diversity of foods.

The most puzzling mammals are the arboreal squirrels. Many or all of them are diurnal and similar to primates in other aspects of behavior and ecology, but their brains seem to have remained at the same rather low level as those of most other rodents. This could be partly due to the fact that they are seldom gregarious, perhaps because they tend to hoard food in caches. Hoarding must favor dispersal of individuals, if only to avoid pilfering.

Granted or assuming that intelligence is positively correlated with the number and diversity of cues to be responded to, it would be nice to know why.

Intelligent behavior can be roughly equated with learning. The alternative to intelligent or learned behavior is the kind of pattern that has been called "instinctive" or "innate," i.e., species-typical and apparently semiautomatic. "Learned" and "innate" patterns may intergrade through many intermediates, but the two extremes are fairly distinct. Both have their good and bad points.

"Innate" reactions tend to be comparatively rapid. They usually are performed with a minimum of preliminary hesitation. Speed of performance probably always is advantageous per se. It allows more time for other necessary or desirable activities. But "innate" reactions also tend to be rather inflexible, and most of them seem to be released by simple stimuli or small numbers of configurations of stimuli. Thus, they are quite likely to be performed mistakenly, inappropriately, when the stimuli are complex or unusual.

Learned reactions may be more flexible and less apt to be performed inappropriately, but they probably will tend to be less rapid on the average. An animal responding to a stimulus in the light of experience alone must spend some significant time, a few seconds, in calculation, comparison, and recall.

This brings up another matter. None of the most intelligent primates, or birds, is very small. It is difficult to imagine how medium to large size could be a direct impulse to intelligence, but it does seem to be a prerequisite. There could be two reasons: frequent and varied learning may not be possible for brains below a certain critical mass, and small species simply cannot get large brains into small heads (see also Gould, 1974); or the partial protection against predators conferred by large size may reduce the disadvantages of slowness of responses.

The size factor may help to explain why some species do not conform to what seem to be the general rules. The smallness of most tree squirrels, whatever its cause, must have contributed to their continuing dullness. Squirrel Monkeys, which are so adaptable and live in such diverse habitats, may have remained less intelligent than capuchins or spider monkeys solely because of their smallness.

Ignoring exceptions, it might be supposed that the increased intelligence of most primates and many parrots and corvids could be related to any one or all of several sets of circumstances. 1. These animals could have unusual amounts of time available in which to make up their minds. 2. Mistakes might be more dangerous or otherwise disadvantageous for them than for most other animals. 3. Their physical and social surroundings might be such that they would be particularly likely to make mistakes if they did not take special precautions.

There is no evidence that the habitats of most primates, parrots, and corvids are either more or less dangerous on the whole than many others, or that the speed of interactions among either individuals or species of other groups in the same environments is noticeably slower than that of the corresponding patterns of related forms of similar sizes living elsewhere. This would seem to leave the third possibility as probably most important.

The complexities of diurnal, arboreal, gregarious life may be so great that semiautomatic responses would be performed in the wrong places or at the wrong times or even fail to "go off" on many occasions, simply because they could be triggered by only a few releasers, or because the relevant stimuli were so varied, and, therefore, unpredictable, that it would be impossible or prohibitively expensive to evolve a sufficient number of "preset" fixed action patterns.

The final argument can be summarized. For most animals, "instinctive" or "innate" responses are most effective. For a few groups of animals, such as many primates, only learned responses are adequate. The habit of learning may be difficult to acquire. Natural selection will favor its appearance or elaboration only in special circumstances. Once acquired, however, it opens opportunities for further development. Intelligence may begin as a *pis aller*, apparently a last resort, but it can also be a preadaptation to new ways of life and greater efficiency and progress in the end.

At least one hopes so.

appendix one
Some Groups of Amazonian Monkeys

In the course of my observations in the upper Amazon, I counted the numbers of the groups of some of the local species of monkeys. These are presented below, in tabular form, for the record. They may be of use to other students in the future.

Table 1. Social Units of *Pithecia monacha* in the Caquetá

No. of Individuals per Unit	No. of Units Counted
1	5
2	9
3	6
4	2
5	1
6	1

The total number of units counted was 24. The total number of individuals counted was 60. The average number of individuals per unit was 2.50.

NOTE: Many more individuals were seen than are included in the table. The units that are recorded here (and in the following tables) are those seen under particularly favorable observational conditions, when I was fairly certain that I had counted all the individuals present at the time. In many other cases, it was obvious that I was seeing only part of a group.

Table 2. Social Units of *Saguinus fuscicollis* in Amazonian Colombia

No. of Individuals per Unit	No. of Units Counted
1	1
2	4
3	4
4	2
5	4
6	5
7	1

The total number of units counted was 21. The total number of individuals counted was 86. The average number of individuals per unit was 4.10.

Two groups of two individuals and one group of three were counted in the Putumayo. The rest were counted in the Caquetá.

Table 3. Social Units of *Saguinus graellsi* near Santa Rosa in the Putumayo

No. of Individuals per Unit	No. of Units Counted
4	3
5	2
6	2
7	1
9	1
12	1

The total number of units counted was 10. The total number of individuals counted was 62. The average number of individuals per unit was 6.20.

appendix two
A Partial "Synoptic" List of Ceboids

It is not yet possible to compile a real synoptic list of living ceboids, but it might be interesting to suggest an approximation thereof, to indicate what would appear to be plausible arrangement in the present state of our knowledge.

There does not seem to be any compelling reason to recognize more than one taxonomic family among ceboids; nor would it be convenient to recognize separate subfamilies or tribes. If any one of the 9 distinctive genera or groups of genera were to be put apart in a higher category, it probably would be necessary to separate all the others at the same level, which would lead to an undesirable proliferation of hierarchical ranks of minimal content.

The sequence of genera and species might be as follows:

Callimico. 1 species, *goeldii.*

Leontopithecus. 1 species, *rosalia.*

Saguinus. Several subgenera and many species. (*Oedipomidas*) with 3 species: *leucopus, oedipus,* and *geoffroyi.* (*Tamarinus*) with 6 species: *fuscicollis, nigricollis, graellsi, mystax, labiatus,* and *imperator.* (*Tamarin*) with 1 species, *midas.* (*Marikina*) with 2 species, *bicolor* and *martinsi.*

Callithrix. 2 or 3 species or species groups: *humeralifer, argentata,* and "*jacchus.*" (The last is a problem. It includes a variety of local forms of distinctive appearance that were supposed to be allopatric. Recently, Coimbra-Filho and Mittermeier, 1973b, have found that several of them seem to overlap very slightly. Perhaps the overlaps are only temporary, or due to human interference.)

Cebuella. 1 species, *pygmaea.*

Aotus. 1 species, *trivirgatus.*

Alouatta. 5 or 6 species: *villosa, seniculus, guariba, belzebul, caraya,* and perhaps *pigra.* (The form called *pigra* is obviously a close relative of the *villosa*, or *palliata,* group. It may, how-ever, be just slightly sympatric with another form of the group, *mexicana,* in Tabasco—see J. D. Smith, op. cit. The problem here is much the same as in the case of *Callithrix* "*jacchus.*")

Pithecia. Probably 4 species: *pithecia, satanas, melanocepha-la,* and *calva.*

Callicebus. 2 or 3 species: *torquatus, moloch,* and perhaps *personatus.*

Lagothrix. 1 species, *lagothricha.*

Ateles. 2 species, *arachnoides* and *paniscus.*

Saimiri. 1 species, *sciureus.*

Cebus. 4 species: *capucinus, nigrivittatus, albifrons,* and *apella.*

Bibliography

Albignac, R. 1973. Mammifères Carnivores. *Faune de Madagascar* 36. Paris. ORSTOM and CNRS.

Alcock, J. 1972. The evolution of the use of tools by feeding animals. *Evolution* 26: 464–473.

Altmann, S. A. 1959. Field observations on a howling monkey society. *J. Mammal.* 40: 317–330.

———. 1962. A field study of the sociobiology of Rhesus Monkeys, *Macaca mulatta. Ann. N.Y. Acad. Sci.* 102: 338–435.

———. 1965. Sociobiology of Rhesus Monkeys. II. Stochastics of social communication. *J. Theor. Biol.* 8: 490–552.

———. 1967. The structure of primate social communication. In *Social communication among primates*, ed. S. A. Altmann. Chicago: Univ. Chicago Press.

———. 1968a. Sociobiology of Rhesus Monkeys. III. The basic communication network. *Behaviour* 32: 17–32.

———. 1968b. Sociobiology of Rhesus Monkeys. IV. Testing Mason's hypothesis of sex differences in affective behavior. *Behaviour* 32: 49–69.

———, and J. Altmann. 1970. *Baboon ecology*. Chicago: Univ. Chicago Press.

Andrew, R. J. 1963. The origin and evolution of the calls and facial expressions of the primates. *Behaviour* 20: 1–109.

———. 1964. The displays of the primates. In *Evolutionary and genetic biology of primates*, vol. 2, ed. J. Buettner-Janusch. New York: Academic Press.

Angst, W. 1974. Das Ausdrucksverhalten des Javaneraffen *Macaca fascicularis* Raffles. *Advances in Ethology*, 15. Berlin and Hamburg: Verlag Paul Parey.

Baldwin, J. D. 1968. The social behavior of adult male Squirrel Monkeys (*Saimiri sciureus*) in a seminatural environment. *Folia Primat.* 9: 281–314.

Baldwin, J. D. 1969. The ontogeny of social behavior of Squirrel Monkeys (*Saimiri sciureus*) in a seminatural environment. *Folia Primat.* 11: 35–79.

———. 1971. The social organization of a semifree-ranging troop of Squirrel Monkeys (*Saimiri sciureus*). *Folia Primat.* 14: 23–50.

———, and J. I. Baldwin. 1971. Squirrel Monkeys (*Saimiri*) in natural habitats in Panama, Colombia, Brazil, and Peru. *Primates* 12: 45–61.

———. 1972a. Population density and use of space in Howling Monkeys (*Alouatta villosa*) in southwestern Panama. *Primates* 13: 371–379.

———. 1972b. The ecology and behavior of Squirrel Monkeys (*Saimiri oerstedii*) in a natural forest in western Panama. *Folia Primat.* 18: 161–184.

———. In press. Primate populations in Chiriquí, Panama. In *Neotropical primates: field studies and conservation.* Washington: NAS.

Baldwin, L. A., and G. Teleki. 1972. Field research on baboons, drills, and Geladas: an historical, geographical, and bibliographical listing. *Primates* 13: 427–432.

———. 1973. Field research on chimpanzees and gorillas: an historical, geographical, and bibliographical listing. *Primates* 14: 315–330.

———. 1974. Field research on gibbons, Siamangs, and Orangutans: an historical, geographical, and bibliographical listing. *Primates* 15: 365–376.

Bates, H. W. 1863. *The naturalist on the River Amazon.* London: John Murray.

Battistini, R., and G. Richard-Vindard, eds. 1972. *Biogeography and ecology of Madagascar.* The Hague: W. Junk.

Bauchot, R., and H. Stephan. 1966. Données nouvelles sur l'encephalisation des insectivores et des prosimiens. *Mammalia* 30: 160–196.

Beard, J. S. 1953. The savanna vegetation of northern tropical America. *Ecol. Monogr.* 23: 149–215.

Beck, B. B. 1973. Cooperative tool use by captive Hamadryas Baboons. *Science* 182: 594–597.

Benirschke, K., R. A. Cooper and V. Desroches. 1974. Cytogenetic observations in South American primates. Paper presented at the 73rd Annual Meeting of the American Anthropological Association, Mexico City.

Bennett, C. F., Jr. 1963. A phytophysiognomic reconnaissance of Barro Colorado Island, Canal Zone. *Smithson. Misc. Coll.* 145 (7): 1–8.

Bernstein, I. S. 1964. A field study of the activities of howler monkeys. *Anim. Behav.* 12: 92–97.

———. 1967a. Intertaxa interactions in a Malayan primate community. *Folia Primat.* 7: 198–207.

———. 1967b. A field study of the Pigtail Monkey (*Macaca nemestrina*). *Primates* 8: 217–228.

———. 1968. The Lutong of Kuala Selangor. *Behaviour* 32: 1–16.

Bertrand, M. 1969. *The behavioral repertoire of the Stumptailed Macaque.* Basel and New York: S. Karger.

Bishop, A. 1962. Control of the hand in lower primates. *Ann. N.Y. Acad. Sci.* 102: 316–337.

———. 1964. Use of the hand in lower primates. In *Evolutionary and genetic biology of primates*, vol. 2, ed. J. Buettner-Janusch. New York: Academic Press.

Booth, A. H. 1957. Observations on the natural history of the Olive Colobus Monkey, *Procolobus verus* (van Beneden). *Proc. Zool. Soc. Lond.* 129: 421–431.

Booth, C. 1962. Some observations on behavior of *Cercopithecus* monkeys. *Ann. N.Y. Acad. Sci.* 102: 477–487.

Bourlière, F., M. Bertrand, and C. Hunkeler, 1969. L'écologie de la Mone de Lowe (*Cercopithecus campbelli lowei*) en Côte d'Ivoire. *Terre et Vie* 1969, 135–163.

Brown, W. L., Jr., and E. O. Wilson. 1956. Character displacement. *Syst. Zool.* 5: 49–64.

Brumback, R. A., and D. O. Willenberg. 1973. Serotaxonomy of *Aotus*. A preliminary study. *Folia Primat.* 20: 106–111.

Buettner-Janusch, J., ed. 1962. The relatives of man: modern studies of the relation of the evolution of nonhuman primates to human evolution. _Ann. N.Y. Acad. Sci._ 102: 181–514.

———. 1966. A problem in evolutionary systematics: nomenclature and classification of baboons, genus _Papio. Folia Primat._ 4: 288–308.

Buxton, A. P. 1952. Observations on the diurnal behavior of the Redtail Monkey (_Cercopithecus ascanius schmidti_ Matschie) in a small forest in Uganda. _J. Anim. Ecol._ 21: 25–58.

Cabrera, A., and J. Yepes. 1940. _Historia natural Ediar; mamíferos sud-americanos._ Buenos Aires: Cia. Argentina de Editores.

Campbell, C.B.G. 1966. The relationships of the tree shrews: the evidence of the nervous system. _Evolution_ 20: 276–281.

Carpenter, C. R. 1934. A field study of the behavior and social relations of howling monkeys (_Alouatta palliata_). _Comp. Psych. Monogr._ 10 (2): 1–168.

———. 1935. Behavior of red spider monkeys in Panama. _J. Mammal._ 16: 171–180.

———. 1940. A field study in Siam of the behavior of social relations of the gibbon (_Hylobates lar_). _Comp. Psych. Monogr._ 16 (5): 1–212.

———. 1962. Field studies of primate populations. In _Roots of behavior_, ed. E. L. Bliss. New York: Harper.

———. 1965. The howlers of Barro Colorado Island. In _Primate behavior_, ed. J. DeVore. New York: Holt.

Cartmill, A. 1974a. Rethinking primate origins. _Science_ 184: 436–443.

———. 1974b. Pads and claws in arboreal locomotion. In _Primate locomotion_, ed. F. A. Jenkins, Jr. New York: Academic Press.

Causey, O. R., H. W. Laemmert, Jr., and G. S. Hayes. 1948. The home range of Brazilian _Cebus_ monkeys in a region of small residual forest. _Amer. J. Hyg._ 47: 304–314.

Chalmers, N. R. 1968a. Group composition, ecology, and daily activities of free living mangabeys in Uganda. *Folia Primat.* 8: 247–262.

———. 1968b. The social behavior of free living mangabeys in Uganda. *Folia Primat.* 8: 263–281.

———. 1968c. The visual and vocal communication of free living mangabeys in Uganda. *Folia Primat.* 9: 258–280.

Chivers, D. 1969. On the daily behavior and spacing of howling monkey groups. *Folia Primat.* 10: 48–102.

———. 1974. *The Siamang in Malaya.* Contributions to Primatology 4. Basel: S. Karger.

Chomsky, N. 1957. *Syntactic structures.* The Hague: Mouton.

———. 1967. The general properties of language. In *Brain mechanisms underlying speech,* ed. F. L. Darley. New York and London: Grune and Stratton.

Christen, A. 1974. Fortpflanzungsbiologie und Verhalten bei *Cebuella pygmaea* und *Tamarin tamarin. Advances in Ethology,* 14. Berlin and Hamburg: Verlag Paul Parey.

Coelhlo, A. M., C. A. Bramblett, L. B. Quick, and S. S. Bramblett. 1974. Energy budgets of howler and spider monkeys in Guatemala: a socio-bioenergetic analysis of population density and resource availability. Paper presented at the 73rd Annual Meeting of the American Anthropological Association, Mexico City.

Coimbra-Filho. A. F. 1971. Os sagüis do gênero. *Callithrix* da regiaõ oriental brasileira e um caso de duplo-hybridismo entre três de suas formas (Callithricidae, Primates). *Rev. Brasil. Biol.* 31: 377–388.

———, and R. A. Mittermeier. 1973a. Distribution and ecology of the genus *Leontopithecus* Lesson, 1940 in Brazil. *Primates* 14: 47–66.

———. 1973b. New data on the taxonomy of the Brazilian marmosets of the genus *Callithrix* Erxleben, 1774. *Folia Primat.* 20: 241–264.

Cole, J. 1963. Macaca nemestrina studied in captivity. *Symp. Zool. Soc. Lond.* 10: 105-114.

Collias, N. E., and C. H. Southwick. 1952. A field study of population density and social organization in howling monkeys. *Proc. Amer. Phil. Soc.* 96: 143–156.

Cooper, R. W. 1968. Squirrel Monkey taxonomy and supply. In *The Squirrel Monkey*, eds. L. A. Rosenblum and R. W. Cooper. New York: Academic Press.

Cramer, D. L. 1968. Anatomy of the thoracic and abdominal viscera. In *Biology of the howler monkey (Alouatta caraya)*, eds. M. R. Malinow. Basel and New York: S. Karger.

Crook, J. H. 1966. Gelada Baboon herd structures and movement: a comparative report. *Symp. Zool. Soc. Lond.* 18: 237–258.

———, and P. Aldrich-Blake. 1968. Ecological and behavioral contrasts between sympatric ground dwelling primates in Ethiopia. *Folia Primat.* 8: 192–227.

———, and J. S. Gartlan. 1966. Evolution of primate societies. *Nature* 210: 1200–1203.

Cruz Lima, E. da. 1945. *Mammals of Amazonia. General introduction and primates.* Contr. Mus. Paraense Emilio Goeldi de Hist. Nat. e Ethnogr. Belem: Oficina Grafica Mauá.

Darlington, P. J. 1957. *Zoogeography: the geographical distribution of animals.* New York: Wiley.

Deag, J. M., and J. H. Crook. 1971. Social behavior and "agnostic buffering" in the wild Barbary Macaque *Macaca sylvana* L. *Folia Primat.* 15: 183–200.

DeBoer, L.E.M. 1974. Cytotaxonomy of the Platyrrhini (Primates). *Genen Pahenen* 17: 1–115.

De Schauensee, R. M. 1970. *A guide to the birds of South America.* Wynnewood, Pa: Livingston.

Devore, I., ed. 1965. *Primate behavior. Field studies of monkeys and apes.* New York: Holt.

Dolhinow, P., ed. 1972. *Primate patterns.* New York: Holt.

D'Orbigny, A. 1836. *Voyage dans l'Amérique méridionale.*

Dumond, F. V. 1968. The Squirrel Monkey in a seminatural

environment. In *The Squirrel Monkey*, eds. L. A. Rosenblum and R. W. Cooper. New York: Academic Press.

Dunbar, R.I.M., and E. P. Dunbar. 1974. Ecology and population dynamics of *Colobus guereza* in Ethiopia. *Folia Primat.* 21: 188–208.

Eisenberg, J. F., and R. E. Kuehn. 1966. The behavior of *Ateles geoffroyi* and related species. *Smithson. Misc. Coll.* 151 (8): 1–63.

——, and P. Leyhausen. 1972. The phylogenesis of predatory behavior in mammals. *Z. Tierpsychol.* 30: 59–93.

——, N. A. Muckenhirn, and R. Rudran. 1972. The relation between ecology and social structure in primates. *Science* 176: 863–874.

——, and R. W. Thorington, Jr. 1973. A preliminary analysis of a Neotropical Mammal Fauna. *Biotropica* 5: 150–161.

Elliot-Smith, G. 1927. *Essays on the evolution of man*, 2nd ed. Oxford: Oxford Univ. Press.

Epple, G. 1967. Vergleichende Untersuchungen über Sexual- und Sozialverhalten der Krallenaffen (Hapalidae). *Folia Primat.* 7: 37–65.

——. 1968. Comparative studies on vocalization in marmoset monkeys (Hapalidae). *Folia Primat.* 8: 1–40.

——, and R. Lorenz. 1967. Vorkommen, Morphologie und Funktion der Sternaldrüse bei den Playtrrhini. *Folia Primat.* 7: 98–126.

Epple-Hösbacher, G. 1967. Soziale Kommunikation bei *Callithrix jacchus* Erxleben, 1777. In *Neue Ergebnisse der Primatologie*, eds. D. Starck, R. Schneider, and H.-J. Kuhn. Stuttgart: Gustav Fischer Verlag.

Erikson, G. E. 1963. Brachiation in the New World monkeys. *Symp. Zool. Soc. Lond.* 10: 135–164.

Ewer, R. F. 1973. *The carnivores.* Ithaca: Cornell Univ. Press.

Fairbanks, L. 1974. An analysis of subgroup structure and process in a captive Squirrel Monkey (*Saimiri sciureus*) colony. *Folia Primat.* 21: 209–224.

Fiedler, W. 1956. Übersicht über das System der Primates. In

Primatologia, vol. 1, eds. H. Hofer, A. H. Schultz, and D. Stark. Basel and New York: S. Karger.

Fooden, J. 1963. A revision of the woolly monkeys (genus *Lagothrix*). *J. Mammal.* 44: 213–247.

————. 1964. Stomach contents and gastro-intestinal proportions in wildshot Guianian monkeys. *Amer. J. Phys. Anthrop.* 22: 227–232.

Freese, C. In press. Censusing *Alouatta palliata, Ateles geoffroyi,* and *Cebus capucinus* in the Costa Rican dry forest. In *Neotropical primates: field studies and conservation.* Washington: NAS.

Gardner, B., and R. A. Gardner. 1971. Two-way communication with an infant chimpanzee. In *Behavior of non-human primates,* eds. A. Schrier and F. Stollnitz, vol. 4. New York: Academic Press.

Gardner, R. A., and B. Gardner. 1969. Teaching sign language to a chimpanzee. *Science.* 165: 664–672.

Gauthier, J. P. 1967. Émissions sonores liées à la cohésion du groupe et aux manifestations d'alarmes dans les bandes de talapoins (*Miopithecus talapoin*). *Biol. Gabon,* 4: 311–329.

————, and A. Gauthier-Hion. 1969. Les associations polyspécifiques chez les Cercopithecidae du Gabon. *Terre et Vie* 23: 164–201.

Gauthier-Hion, A. 1966. L'écologie et l'éthologie du talapoin (*Miopithecus talapoin*). *Biol. Gabon.* 4: 311–329.

————, and J.-P. Gauthier. 1974. Les associations polyspécifiques de cercopithèques du Plateau de M'passa (Gabon). *Folia Primat.* 22: 134–177.

Goldman, E. A. 1920. Mammals of Panama. *Smithson. Misc. Coll.* 69 (5): 1–309.

Gould, S. J. 1974. Sizing up human intelligence. *Natural History* 83: 10–14.

Goulissachvili, V. Z. 1964. Les savanes des pays tropicaux et subtropicaux. In *The ecology of man in the tropical environment.* Morges, Switzerland: IUCN.

Haddow, A. J. 1952. Field and laboratory studies on an African monkey, *Cercopithecus ascanius schmidti* Matschie. *Proc. Zool. Soc. Lond.* 122: 22–26.

Haffer, J. 1967. Speciation in Colombian forest birds. *Amer. Mus. Novit.* 2294.

———. 1969. Speciation in Amazonian forest birds. *Science* 165: 131–137.

———. 1974. Avian speciation in tropical South America. *Publ. Nuttall Orn. Club* 14.

Hall, K.R.L. 1965. Behavior and ecology of the wild Patas Monkey, *Erythrocebus patas* in Uganda. *J. Zool.* 148: 15–87.

———, and J. S. Gartlan. 1965. Ecology and behavior of the Vervet Monkey (*Cercopithecus aethiops*), Lolui Island, Lake Victoria. *Proc. Zool. Soc. Lond.* 145: 37–56.

Hardin, G. 1960. The competitive exclusion principle. *Science* 131: 1292–1297.

Heltne, P. G., D. C. Turner, and N. Scott. In press. Comparison of census data on *Alouatta villosa* (= palliata) from Costa Rica and Panama. In *Neotropical primates: field studies and conservation*. Washington: NAS.

Hernández-Camacho, J., and R. W. Cooper. In press. The nonhuman primates of Colombia. In *Neotropical primates: field studies and conservation*. Washington. NAS.

Hershkovitz, P. 1949. Mammals of northern Colombia. Preliminary report no. 4: monkeys (Primates), with taxonomic revisions of some forms. *Proc. U. S. Nat. Mus.* 98: 323–347.

———. 1958. A geographic classification of neotropical mammals. *Fieldiana, zoology* 36: 581–620.

———. 1963. A systematic and zoogeographic account of the monkeys of the genus *Callicebus* (Cebidae) of the Amazonas and Orinoco river basins. *Mammalia* 27: 1–79.

———. 1966. Taxonomic notes on tamarins, genus *Saguinus* (Callithricidae, Primates), with descriptions of four new forms. *Folia Primat.* 381–395.

Hershkovitz, P. 1968. Metachromism or the principle of evolutionary change in mammalian tegumentary colors. *Evolution* 22: 556–575.

———. 1969. The recent mammals of the neotropical region: a zoogeographic and ecological review. *Quart. Rev. Biol.* 44: 1–70.

———. 1970. Notes on Tertiary platyrrhine monkeys and description of a new genus from the late Miocene of Colombia. *Folia Primat.* 12: 1–37.

———. 1972. The recent mammals of the neotropical region: a zoogeographic and ecological review. In *Evolution, mammals, and southern continents,* eds. A. Keast, F. C. Erk, and B. Glass. Albany: SUNY Press.

Hewes, G. W. 1973. Primate communication and the gestural origin of language. *Current Anthropology* 14: 5–24.

Hill, W.C.O. 1953. *Primates. I. Strepsirhini.* Edinburgh: Univ. Edinburgh Press.

———. 1957. *Primates. III. Pithecoidea Platyrrhini.* Edinburgh: Univ. Edinburgh Press.

———. 1960. *Primates. IV. Cebidae, Part A.* Edinburgh: Univ. Edinburgh Press.

———. 1962. *Primates. V. Cebidae, Part B.* Edinburgh: Univ. Edinburgh Press.

Hladik, A., and C. M. Hladik. 1969. Rapports trophiques entre végétation et primates dans la forêt de Barro Colorado (Panama). *Terre et Vie* 1969: 25–117.

Hladik, C. M., and A. Hladik. 1967. Observations sur le rôle des primates dans la dissemination des végétaux de la forêt gabonaise. *Biol. Gabon.* 3: 43–58.

———. 1972. Disponibilités alimentaires et domaines vitaux des primates a Ceylan. *Terre et Vie* 2: 149–215.

———, J. Bousset, P. Valdebouze, G. Viroben, and J. Delort-Laval. 1971. Le régime alimentaire des primates de l'île de Barro Colorado (Panama). *Folia Primat.* 16: 85–122.

Hladik, C. M., and P. Charles-Dominique. 1971. Lépilémur et autres lémuriens du sud de Madagascar. *Science et Nature* 106: 30–38.

Hockett, C. F. 1959. Animal "languages" and human language. In *The evolution of man's capacity for language*, ed. J. N. Spuhler. Detroit: Wayne Univ. Press.

——. 1960. Logical considerations in the study of animal communication. In *Animal sounds and communication*, eds. W. E. Lanyon and W. N. Tavolga. Washington: *Amer. Inst. Biol. Sci.*

Hoffstetter, R. 1972. Relationships, origins, and history of the ceboid monkeys and caviomorph rodents: a modern reinterpretation. In *Evolutionary Biology*, eds. T. Dobzhansky, M. K. Hecht, and W. C. Steere. Vol. 6. New York: Appleton-Century-Crofts.

Humboldt, A. de, and A. Bonpland. 1811-1812. Recueil d'observations de zoologie et d'anatomie comparée faites dans l'océan atlantique, dans l'intérieur du nouveau continent et dans la mer du sud pendant les années 1799, 1800, 1801, 1802 et 1803.

Hunkeler, C., F. Bourlière, and M. Bertrand. 1972. Le comportement social de la Mone de Lowe (*Cercopithecus campbelli lowei*). *Folia Primat.* 17: 218–236.

Husson, A. M. 1957. Notes on the primates of Suriname. *Studies on the fauna of Suriname and other guyanas* 2: 13–40.

Hutchinson, G. E. 1959. Why are there so many kinds of animals? *Amer. Naturalist* 93: 145–159.

——. 1963. Natural selection, social organization, hairlessness, and the australopithecine canine. *Evolution* 17: 588–589.

Huxley, T. H. 1876. *Man's place in nature.* New York: Appleton.

Imanishi, K., and S. A. Altmann. 1965. *Japanese monkeys.* Published by the editor (S. A. Altmann), Univ. Alberta.

Jay, P., ed. 1968. *Primates. Studies in adaptation and variability.* New York: Holt.

Jerison, H. I. 1973. *Evolution of the brain and intelligence.* New York: Academic Press.

Jolly, A. 1966. *Lemur behavior.* Chicago: Univ. Chicago Press.

———. 1972. *The evolution of primate behavior.* New York: Macmillan.

Jones, C., and J. Sabater Pi. 1968. Comparative ecology of *Cercocebus albigena* (Gray) and *Cercocebus torquatus* (Kerr) in Rio Muni, West Africa. *Folia Primat.* 9: 99–113.

Kaufmann, J. H. 1965. Studies on the behavior of captive tree shrews (*Tupaia glis*). *Folia Primat.* 3: 50–74.

Keast, A. 1972a. Introduction: the southern continents as backgrounds for mammalian evolution. In *Evolution, mammals, and southern continents,* eds. A. Keast, F. C. Erk, and B. Glass. Albany: SUNY Press.

———. 1972b. Comparisons of contemporary mammal faunas of southern continents. In *Evolution, mammals, and southern continents,* eds. A. Keast, F. C. Erk, and B. Glass. Albany: SUNY Press.

———. 1972c. Australian mammals: zoogeography and evolution. In *Evolution, mammals, and southern continents,* eds. A. Keast, F. C. Erk, and B. Glass. Albany: SUNY Press.

Kellogg, R., and E. A. Goldmann. 1944. Review of the spider monkeys. *Proc. U.S. Nat. Mus.* 96: 1–45.

Kinzey, W. G. 1972. Canine teeth of the monkey *Callicebus moloch*: lack of sexual dimorphism. *Primates* 13: 365–369.

Klein, L. L. 1972. The ecology and social organization of the spider monkey, *Ateles belzebuth.* Ph.D. Thesis, Univ. Calif., Berkeley.

———, and D. J. Klein. In press. Neotropical primates: aspects of habitat usage, population density, and regional distribution in La Macarena, Colombia. In *Neotropical primates: field studies and conservation.* Washington: NAS.

Klüver, H. 1933. *Behavior mechanisms in monkeys.* Chicago: Univ. Chicago Press.

Kortlandt, A. 1967. Reply to "More on tool-using among primates." *Current Anthropology* 8: 253.

———. 1974. New perspectives on ape and human evolution. *Current Anthropology* 15: 427–430.

———, and M. Kooij. 1963. Protohominid behavior in primates (preliminary communication). In The Primates, *Symp. Zool. Soc. Lond.* 10.

Krieg, H. 1930. Biologische Reisestudien in Südamerika. XVI. Die Affen des gran Chaco und seiner Grenzgebiete. *Z. Morph. u. Ökol.* 18: 760–785.

Kruuk, H. 1972. *The Spotted Hyena.* Chicago: Univ. Chicago Press.

Kühlhorn, F. 1939. Beobachtungen über das Verhalten von Kapuzineraffen in freier Wildbahn. *Z. f. Tierpsych.* 3: 147–151.

———. 1943. Beobachtungen über der Biologie von *Cebus apella. Zool. Gart.* 15: 221–234.

Kummer, H. 1968. *Social organization of Hamadryas Baboons.* Chicago: Univ. Chicago Press.

———. 1971. *Primate societies.* Chicago and New York: Aldine Atherton.

Lawick-Goodall, J. van. 1968. The behavior of free-living chimpanzees in the Gomba Stream Reserve. *Anim. Behav. Monogr.* 1 (3): 161–311.

———. 1970. Tool-using in primates and other vertebrates. In *Advances in the study of behavior,* 3, eds. D. S. Lehrman, R. A. Hinde, and E. Shaw. New York: Academic Press.

———, and H. van Lawick. 1970. *Innocent killers.* London: Collins.

Lehmann V., F. C. 1969. El Aguila Real de montaña (*Oroaetes isidori* Des Murs). Contribuciones al estudio de la fauna de Colombia XIV. *Bol. Acad. Hist. del Valle del Cauca* 150: 3–29.

Leigh, E. G., Jr. In press. Structure and climate in tropical rain forest.

Lenneberg, E. H. 1964. The capacity for language acquisition. In *The structure of language*, eds. J. A. Fodor and J. J. Katz. Englewood, N.J.: Prentice-Hall.

———. 1967. *Biological foundation of language*. New York: Wiley.

Le Roux, G. 1967. Contribution à l'étude des moyens d'intercommunication chez le Ouistiti à Pinceaux (*Hapale jacchus*). Univ. Rennes: Diplôme d'étude Apprefondie.

Locker Pope, B. 1968. Population characteristics. In *Biology of the howler monkey (Alouatta caraya)*. Bibl. Primat. 7. Basel and New York: S. Karger.

Lorenz, K. 1952. *King Solomon's ring*. New York: Crowell.

———. 1966. *On aggression*. London: Methuen.

Lorenz, R. 1971. Goeldi's monkey *Callimico goeldii* Thomas preying on snakes. *Folia Primat.* 15: 133–142.

MacArthur, R. H. 1972. *Geographical ecology*. New York: Harper and Row.

Mackinnon, J. 1974. The behaviour and ecology of wild Orang-utans (*Pongoe pygmaeus*). *Anim. Behav.* 22: 3–74.

Malinow, M. R. 1968. Introduction. In *Biology of the howler monkey (Alouatta caraya)*. Bibl. Primat. 7. Basel and New York: S. Karger.

Marler, P. 1970. Vocalizations of East African monkeys. I. Red Colobus. *Folia Primat.* 13: 81–91.

———. 1972. Vocalizations of East African monkeys. II. Black and White Colobus. *Behaviour* 42: 175–197.

———. 1973. A comparison of vocalizations of Red-tailed Monkeys and Blue Monkeys, *Cercopithecus ascanius* and *C. mitis*, in Uganda. *Z. Tierpsychol.* 33: 223–247.

———, and W. J. Hamilton, III. 1967. *Mechanisms of animal behavior*. New York: Wiley.

Martin, R. D. 1968a. Towards a definition of primates. *Man* 3: 377–401.

———. 1968b. Reproduction and ontogeny in tree-shrews (*Tupaia belangeri*) with reference to their general behavior and taxonomic relationships. *Z. f. Tierpsych.* 25: 409–495, 505–532.

———. 1972. A preliminary study of the Lesser Mouse Lemur (*Microcebus murinus* J. F. Miller 1777). *Advances in Ethology* 9: 43–89.

Mason, W. A. 1966. Social organization of the South American monkey *Callicebus moloch. Tulane Stud. in Zool.* 13: 23–28.

———. 1968. Use of space by *Callicebus* groups. In *Primates. Studies in adaptation and variability,* ed. P. Jay. New York: Holt.

———. 1974. Comparative studies of social behavior in *Callicebus* and *Saimiri*: behavior of male-female pairs. *Folia Primat.* 22: 1–8.

May, R. M. 1972. Will a large complex system be stable? *Nature* 238: 413–414.

———. 1973. *Stability and complexity in model ecosystems,* Monographs in population biology, 6. Princeton: Princeton Univ. Press.

Mayr, E. 1969. Bird speciation in the tropics. *Bull. J. Linn. Soc.* 1: 1–17.

McKenna, M. C. 1966. Paleontology and the origin of primates. *Folia Primat.* 4: 1–25.

Michael, R. P., and J. H. Crook, eds. 1973. *Comparative ecology and behaviour of primates.* New York: Academic Press.

Mittermeier, R. A. 1973. Group activity and population dynamics of the howler monkey on Barro Colorado Island. *Primates* 14: 1–19.

———, and J. G. Fleagle. In press. The locomotor repertoires of *Ateles geoffroyi* and *Colobus guereza.*

Mivart, St. G. J. 1873. On *Lepilemur* and *Cheirogaleus* and on the zoological rank of the Lemuroidea. *Proc. Zool. Soc. Lond.*: 484–510.

Montgomery, G. G., and M. E. Sunquist. In press. Impact of sloths on neotropical forest energy flow and nutrient cycling. In *Trends in tropical ecology: Ecological Studies IV*, eds. E. Medina and F. Golley. New York: Springer Verlag.

Moreau, R. E. 1966. *The bird faunas of Africa and its islands.* New York: Academic Press.

Moynihan, M. 1962. The organization and probable evolution of some mixed species flocks of neotropical birds. *Smithson. Misc. Coll.* 143 (7): 1–140.

————. 1964. Some behavior patterns of platyrrhine monkeys. I. The Night Monkey (*Aotus trivirgatus*). *Smithson. Misc. Coll.* 146 (5): 1–84.

————. 1966. Communication in *Callicebus. J. Zool. Lond.* 150: 77–127.

————. 1967. Comparative aspects of communication in New World primates. In *Primate ethology,* ed. D. Morris. London: Weidenfeld & Nicolson.

————. 1968. Social mimicry; character convergence versus character displacement. *Evolution* 22: 315–331.

————. 1970a. The control, suppression, decay, disappearance, and replacement of displays. *J. Theor. Biol.* 29: 85–112.

————. 1970b. Some behavior patterns of platyrrhine monkeys. II. *Saguinus geoffroyi* and some other tamarins. *Smithson. Contr. Zool.* 28: 1–77.

————. 1971. Successes and failures of tropical mammals and birds. *Amer. Naturalist* 105: 371–383.

————. 1973. The evolution of behavior and the role of behavior in evolution. *Breviora* 415: 1–29.

————. In press. Notes on the behavior and ecology of the Pygmy Marmoset, *Cebuella pygmaea,* in Amazonian Colombia. In *Neotropical primates: field studies and conservation.* Washington: NAS.

Muckenhirn, N. A. 1967. The behavior and vocal repertoire of *Saguinus oedipus* (Hershkovitz 1966) (Callithricidae, Primates). Unpubl. M.S. Thesis, Univ. Maryland.

Munemi, K., and M. Tetsuzo. 1972. Ecology and behavior of

the wild Proboscis Monkey *Nasalis larvatus* (Wurmb) in Sabah, Malaysia. *Primates* 13: 213–228.

Napier, J. R., and P. H. Napier. 1967. *A handbook of living primates*. New York: Academic Press.

———, ed. 1970. *Old World Monkeys. Evolution, systemics, and behavior*. New York: Academic Press.

———, and A. C. Walker. 1967. Vertical clinging and leaping—a newly recognized category of locomotor behavior of primates. *Folia Primat.* 6: 204–291.

Neville, M. 1972a. The population structure of Red Howler Monkeys (*Alouatta seniculus*) in Trinidad and Venezuela. *Folia Primat.* 17: 56–86.

———. 1972b. Social relations within troops of Red Howler Monkeys (*Alouatta seniculus*). *Folia Primat.* 18: 47–77.

———. In press. The population and conservation of howler monkeys in Venezuela and Trinidad. *Neotropical primates: field studies and conservation*. Washington: NAS.

Nolte, A. 1955. Field observations on the daily routine and social behavior of common Indian monkeys with special reference to the Bonnet Monkey (*Macaca radiata* Geoffroy). *J. Bombay Nat. Hist. Soc.* 53: 177–184.

Nottebohm, F. 1970. Ontogeny of bird song. *Science* 167, 3920: 950–956.

Oppenheimer, J. R. 1968. Behavior and ecology of the White-faced Monkey, *Cebus capucinus*, on Barro Colorado Island, Canal Zone. Ph.D. Thesis, Univ. Illinois, Urbana.

———. 1973. Social and communicatory behavior in the *Cebus* monkey. In *Behavioral regulators of behavior in primates*, ed. C. R. Carpenter. Lewisburg, Pa.: Bucknell Univ. Press.

———, and E. C. Oppenheimer. 1973. Preliminary observations of *Cebus nigrivittatus* (Primates: Cebidae) on the Venezuelan Llanos. *Folia Primat.* 19: 409–436.

Parsons, R. F., and D. G. Cameron. 1974. Maximum plant species diversity in terrestrial communities. *Biotropica* 6: 202–203.

Patterson, B., and R. Pascual. 1968. The fossil mammal fauna of South America. *Quart. Rev. Biol.* 43: 409–451.

———. 1972. The fossil mammal fauna of South America. In *Evolution, mammals, and southern continents,* eds. A. Keast, F. C. Erk, and B. Glass. Albany: SUNY Press.

Paulian, R. 1961. *La zoogéographie de Madagascar et des îles voisines.* Faune de Madagascar XIII. Tananarive-Tsimbazzaza: Institut de Recherche Scientifique.

Perez-Arbelaez, E. 1956. *Plantas útiles de Colombia.* Bogotá: Camacho Roldan.

Petter, J.-J. 1962a. Recherches sur l'écologie et l'éthologie des lémuriens malgaches. *Mem. Mus. Nat. Hist. Nat.,* nouvelle serie (A), 27 (1): 1–146.

———. 1962b. Ecological and behavioral studies of Madagascar lemurs in the field. *Ann. N.Y. Acad. Sci.* 102: 267–281.

———. 1965. The lemurs of Madagascar. In *Primate behavior,* ed. I. DeVore. New York: Holt.

———. 1972. Order of primates: sub-order of lemurs. In *Biogeography and ecology of Madagascar,* eds. R. Battistini and G. Richard-Vindard. The Hague: W. Junk.

———, and G. Pariente. 1971. Les Indridés Malgaches. *Science et Nature* 106: 15–24.

Ploog, D. W., and P. D. Maclean. 1963. Display of penile erection in Squirrel Monkey (*Saimiri sciureus*). *Anim. Behav.* 11: 32–39.

Poirier, F. E. 1969. The Nilgiri Langur (*Presbytis johnii*) troop: its composition, structure, function and change. *Folia Primat.* 10: 20–47.

———. 1970. The communication matrix of the Nilgiri Langur (*Presbytis johnii*) of South India. *Folia Primat.* 13: 92–136.

———, ed. 1972. *Primate socialization.* New York: Random House.

Pola, Y. V., and C. T. Snowdon. In press. The behavioral ontogeny and vocalizations of Pygmy Marmosets. *Anim. Behav.*

Polidora, V. J. 1964. Learning abilities of New World monkeys. *Amer. J. Phys. Anthrop.* 22: 245–252.

Premack, A. J., and D. Premack. 1972. Teaching language to an ape. *Scientific American* 227: 92–99.

Premack, D. 1971. Language in chimpanzee? *Science* 172: 808–822.

Radinsky, L. B. 1970. The fossil evidence of prosimian brain evolution. In *Advances in Primatology*. I, eds. C. R. Noback and W. Montagna. New York: Appleton-Century-Crofts.

Rahaman, H., and M. D. Parthasarathy. 1969. Studies on the social behaviour of Bonnet Monkeys. *Primates* 10: 149–162.

Ramirez, M., A. L. Rosenberger, and W. G. Kinzey. In press. Vertical clinging and leaping in a neotropical anthropoid.

Rand, A. L. 1936. The distribution and habits of Madagascar birds. *Bull. Amer. Mus. Nat. Hist.* 72: 143–499.

Raven, P. H., and D. I. Axelrod. 1974. Angiosperm biogeography and past continental movements. *Ann. Missouri Botan. Gard.* 61: 539–673.

Remane, A. 1956. Paläontologie und evolution der Primaten. In *Primatologia*, vol. 1, eds. H. Hofer, A. H. Schultz, and D. Stark. Basel and New York: S. Karger.

Rensch, B. 1956. Increase of learning capability by increase of brain size. *Amer. Naturalist* 90: 81–95.

———. 1960. *Evolution above the species level*. New York: Columbia Univ. Press.

Reynolds, V. 1967. *The apes*. London: Cassell.

———, and F. Reynolds. 1965. Chimpanzees of the Budongo Forest. In *Primate behavior*, ed. I. DeVore. New York: Holt.

Richard, A. 1970. A comparative study of the activity patterns and behavior of *Alouatta villosa* and *Ateles geoffroyi*. *Folia Primat.* 12: 241–263.

———. 1974. Intraspecific variation in the social organization and ecology of *Propithecus verreauxi. Folia Primat.* 22: 178–207.

Richards, P. W. 1952. *The tropical rain forest.* Cambridge: Cambridge Univ. Press.

———. 1973. Africa, the "odd man out." In *Tropical forest ecosystems in Africa and South America,* eds. B. J. Meggers, E. S. Ayensu, and W. D. Duckworth. Washington: Smithsonian.

———. 1973b. The tropical rain forest. *Scientific American,* December 1973: 58–67.

Robinson, M. H. 1973. Insect anti-predator adaptations and the behavior of predatory primates. *Act. IV Congr. Latin Zool.* (1970): 811–836.

Rodman, P. S. 1973. Population composition and adaptive organization among orang-utans of the Kutai Reserve. In *Comparative ecology and behaviour of primates,* eds. J. H. Crook and R. P. Michael. London: Academic Press.

Rowell, T. E. 1972. The social behavior of monkeys. Baltimore: Penguin Books.

———. 1973. Social organization of wild Talapoin Monkeys. *Amer. J. Phys. Anthrop.* 38: 593–597.

Rumbaugh, D. M., T. V. Gill, and E. C. von Glasersfeld. 1973. Reading and sentence completion by a chimpanzee (*Pan*). *Science* 182: 731–733.

Sabater Pi, J. 1974. An elementary industry of the chimpanzees in the Okorobiké Mountains, Rio Muni (Republic of Equatorial Guinea), West Africa. *Primates* 15: 351–364.

Sanderson, I. T. 1957. *The monkey kingdom.* New York: Hanover House.

Schaller, G. B. 1963. *The Mountain Gorilla.* Chicago: Univ. Chicago Press.

———. 1972. *The Serengeti Lion.* Chicago: Univ. Chicago Press.

Simons, E. L. 1963. A critical reappraisal of Tertiary primates. In *Evolutionary and genetic biology of primates,* vol. 1, ed. J. Buettner-Janusch. New York: Academic Press.

———. 1964. The early relatives of man. *Scientific American* 211: 50–62.

———. 1972. *Primate evolution*. New York: Macmillan.

———, and D. R. Pilbeam. 1965. Preliminary revision of the Dryopithecinae (Pongidae, Anthropoidea). *Folia Primat.* 3: 81–152.

Simpson, G. G. 1945. The principles of classification and the classification of mammals. *Bull. Amer. Mus. Nat. Hist.* 85: 1–350.

———. 1969. *Biology and man*. New York: Harcourt.

Simpson Vuilleumier, B. 1971. Pleistocene changes in the fauna and flora of South America. *Science* 173: 771–780.

Smith, J. D. 1970. The systematic position of the black howler monkey, *Alouatta pigra* Lawrence. *J. Mammal.* 51: 358–369.

Sorenson, M. W., and C. H. Conaway. 1966. Observations on the social behaviour of tree shrews in captivity. *Folia Primat.* 4: 124–145.

Stephan, H. 1967. Quantitative Vergleiche zur phylogenetischen Entwicklung des Gehirns der Primaten mit Hilfe von Progression-indices. *Mitt. Max-Planck-Gessells.* 2: 63–86.

———. 1972. Evolution of primate brains: a comparative anatomical investigation. In *The functional and evolutionary biology of primates,* ed. R. Tuttle. Chicago and New York: Aldine Atherton.

Stern, J. T., Jr. 1971. Functional myology of the hip and thigh of cebid monkeys and its implications for the evolution of erect posture. *Bibliotheca Primatologica* 14. New York: S. Karger.

———, and C. E. Oxnard. 1973. Primate locomotion: some links with evolution and morphology. *Primatologia* 4, 11: 1–93.

Stirton, R. A. 1951. Ceboid monkeys from the Miocene of Colombia. *Univ. Calif. Publ., Bull. Dept. Geol. Sci.* 28: 315–356.

Struhsaker, T. T. 1967. Behavior of Vervet Monkeys (*Cercopithecus aethiops*). Univ. Calif. Publ. Zool. 82.

——. 1969. Correlates of ecology and social organization among African cercopithecines. *Folia Primat.* 11: 80–118.

Sugiyama, Y. 1965. Behavioral development and social structure in two troops of Hanuman Langurs (*Presbytis entellus*). *Primates* 5: 7–38.

——. 1973. Social organization of wild chimpanzees. In *Behavioral regulators of behavior in primates*, ed. C. R. Carpenter. Lewisburg, Pa.: Bucknell Univ. Press.

Szalay, F. S. 1967. The beginnings of primates. *Evolution* 22: 19–36.

Talbot, L. M. 1964. The biological productivity of the tropical savanna ecosystem. In *The ecology of man in the tropical environment*. Morges, Switzerland: IUCN.

Tappen, N. C. 1960. African monkey distribution. *Current Anthropology* 1: 91–120.

——. 1968. Problems of distribution and adaptation of the African monkeys. In *Taxonomy and phylogeny of Old World primates with references to the origin of man*. Turin: Rosenberg & Sellier.

Tattersall, I. 1969a. Ecology of North Indian *Ramapithecus*. *Nature* 221: 451–452.

——. 1969b. More on the ecology of North Indian *Ramapithecus*. *Nature* 224: 821–822.

Tenaza, R. R., and W. J. Hamilton, III. 1971. Preliminary observations of the Mentawai Islands Gibbon, *Hylobates klossii*. *Folia Primat.* 15: 201–211.

Thorington, R. W., Jr. 1967. Feeding and activity of *Cebus* and *Saimiri* in a Colombian forest. In *Progress in primatology*, eds. D. Starck, R. Schneider, H.-J. Kuhn. Stuttgart: Gustav Fischer.

——. 1968a. Observations on the tamarin *Saguinus midas*. *Folia Primat.* 9: 95–98.

————. 1968b. Observations of Squirrel Monkeys in a Colombian forest. In *The Squirrel Monkey*, eds. L. A. Rosenblum and R. W. Cooper. New York: Academic Press.

————, N. A. Muckenhirn, and G. G. Montgomery. In press. Movements of a wild Night Monkey, *Aotus trivirgatus*. In *Neotropical primates: field studies and conservation*. Washington: NAS.

Thorpe, W. H. 1956. *Learning and instinct in animals*. London: Methuen.

Tokuda, K. 1969. Group size and vertical distribution of New World monkeys in the basin of the Putumayo, the upper Amazon. *Proc. 8th Int. Congr. Anthrop. and Ethnol. Sci.* I: 260–261. Tokyo: Science Council of Japan.

Tuttle, R., ed. 1972. *The functional and evolutionary biology of primates*. Chicago and New York: Aldine Atherton.

Tyndale-Biscoe, H. 1973. *Life of marsupials*. New York: American Elsevier.

Ullrich, W. 1961. Zur Biologie und Soziologie der Colobusaffen (*Colobus guereza caudasus* Thomas 1885). *Zool. Gart.* 25: 305–368.

Van Hoof, J.A.H. 1962. Facial expressions in higher primates. *Symp. Zool. Soc. Lond.* 8: 97–125.

Van Valen, L. 1965. Treeshrews, primates and fossils. *Evolution* 19: 137–151.

————, and R. E. Sloan. 1965. The earliest primates. *Science* 150: 743–745.

Vanzolini, P. E. 1973. Paleoclimates, relief, and species multiplication in equatorial forests. In *Tropical forest ecosystems in Africa and South America: a comparative review*, eds. B. J. Meggers, E. S. Ayensu, and W. D. Duckworth. Washington: Smithsonian.

————, and E. E. Williams. 1970. South American anoles: the geographic differentiation and evolution of the *Anolis chrysolepis* species group (Sauria, Iguanidae). *Arq. Zool. S. Paulo* 19: 1–240.

Vesey-Fitzgerald, L.D.E.F. 1964. Grasslands. In *The ecology of man in the tropical environment*. Morges, Switzerland: IUCN.

Wagner, H. O. 1956. Freilandbeobachtungen an Klammeraffen. *Z. Tierpsychol.* 13: 302–313.

Walker, A. 1967a. Locomotor adaptations in recent and fossil Madagascan lemurs. Ph.D. Thesis. Univ. London, London.

——. 1967b. Patterns of extinction among the subfossil Madagascan lemuroids. In *Pleistocene extinctions,* eds. P. S. Martin and H. E. Wright, Jr. New Haven: Yale Univ. Press.

——. 1974. Locomotor adaptations in past and present prosimian primates. In *Primate locomotion,* ed. F. A. Jenkins, Jr. New York: Academic Press.

Walls, G. L. 1942. The vertebrate eye. *Cranbrook Inst. Sci. Bull.* 19.

Walsh, J., and R. Gannon. 1967. *Time is short and the water rises.* Camden, N. J.: Nelson.

Wetmore, A. 1965. The birds of the republic of Panama. Part I. Tinamidae (tinamous) to Rynchopidae (skimmers). *Smithson. Misc. Coll.* 150.

Wickler, W. 1967. Sociosexual signals and their intraspecific imitation among primates. In *Primate ethology,* ed. D. Morris. London: Weidenfeld and Nicolson.

Williams, E. E., and K. F. Koopman. 1952. West Indian fossil monkeys. *Amer. Mus. Novit.* 1546.

Williams, L. 1969. *Man and monkey.* London: Panther.

Zapfe, H. 1963. Lebensbild von *Megaladapis edwardsi* (Grandidier). *Folia Primat.* 5: 178–187.

Zuckerman, S. 1932. *The social life of monkeys and apes.* London: Routledge & Kegan Paul.

Index

acoustic signals (vocalizations), 157ff; adaptive significance of differences in pitch, 168; discrete vs. intergrading patterns, 170; frequencies of performance, 169; heterogeneous vocalizations of medium pitch, 168; high pitched vocalizations, 157ff; low pitched vocalizations, 157ff; ontogeny, 171; precision, 171; structure of notes, 163; typical basic repertory of New World primates, 157

Aepyornithidae, 197

Africa: climate, 14; history, 14; comparisons with other tropical regions, 208; primate fauna, 198

African Hunting Dog (*Lycaon pictus*), 181

aggression, advantages and disadvantages of, 120; correlation with terrestrial life in open country, 120

Albignac, R., 200

Alcock, J., 180

Aldrich-Blake, P., 201

Alibertia edulis, 104

Allenopithecus, 202

Allocebus, 192

Alouatta, 9, 49, 66, 73, 104, 140, 195, 205, 212; appearance, 55; brain, 63; classification, 54, 223; correlation between vocal signals and visual stimuli, 58; ecologies and distributions, 49ff, 54; extreme specialization, 63; foods and feeding habits, 62f; habitat, 59, 61; hyoid apparatus and loudness of vocalizations, 56; locomotion, 59;

manipulative ability, 59; peculiarities of signal repertory, 163, 168f; phylogenetic relationships, 49; population densities and fluctuations, 64f; reactions to predators, 63; sleeping habits, 63; visual communication, 156f

Alouatta caraya, 55, 63; appearance, 55; population, 65; sex ratio, 112; social groups, 112

Alouatta palliata, see A. villosa

Alouatta seniculus, 58, 60, 132, 163; appearance, 55; feeding, 62; habitat, 69; locomotion, 61; populations, 65; sexual display, 153; social groups, 114

Alouatta villosa, 20, 54, 57, 59, 84, 98, 105, 116, 118, 123, 127, 141, 142, 153, 163, 175, 213; appearance, 55; distribution, 55; feeding, 62; hostile display, 56; interactions with other species, 124ff; locomotion, 61; parasites, 61; partly hostile display by infant, 58; populations, 64; sex ratio, 112; sexual display, 153; social groups, 112; tempo, 63; territorial behavior, contrast with *Callicebus moloch*, 78

Alouatta (villosa) pigra, 55, 62

Altmann, J., 201

Altmann, S. A., 55, 173, 201

Amapá, 36

Amazon basin, 16, 25, 27, 29, 42, 65, 83, 84, 132ff, 135, 138, 144, 222

Anacardium excelsum, 62, 104

Andrew, R. J., 152, 192, 201

Angst, W., 201

Anona spraguei, 93, 104

anteaters (Myrmecophagidae), 16

Anthropoidea, 5

Library of Congress Cataloging in Publication Data

Moynihan, M.
  The New World primates.

  Bibliography: p.
  Includes index.
  1. Cebidae—Behavior. 2. Callithricidae—Behavior.
3. Mammals—Latin America. I. Title.
QL737.P92M68   599'.82   75-3467
ISBN 0-691-08168-9
ISBN 0-691-08169-7 pbk.